4/06

49.95

I0428760

Genetic Engineering

Recent titles in

Historical Guides to Controversial Issues in America
Ron Chepesiuk, Series Editor

Gun Control and Gun Rights
Constance Emerson Crooker

The Pro-Life/Choice Debate
Mark Y. Herring

Genetic Engineering

Mark Y. Herring

Historical Guides to Controversial Issues in America

GREENWOOD PRESS
Westport, Connecticut • London

Library of Congress Cataloging-in-Publication Data

Herring, Mark Youngblood, 1952–
    Genetic engineering / Mark Y. Herring.
        p. cm. — (Historical guides to controversial issues in America)
    Includes bibliographical references and index.
    ISBN 0–313–33473–0
    1. Genetic engineering—Popular works. I. Title. II. Series.
QH442.H49 2006
660.6′5—dc22   2005026304

British Library Cataloguing in Publication Data is available.

Copyright © 2006 by Mark Y. Herring

All rights reserved. No portion of this book may be
reproduced, by any process or technique, without the
express written consent of the publisher.

Library of Congress Catalog Card Number: 2005026304
ISBN: 0–313–33473–0
ISSN: 1541–0021

First published in 2006

Greenwood Press, 88 Post Road West, Westport, CT 06881
An imprint of Greenwood Publishing Group, Inc.
www.greenwood.com

Printed in the United States of America

The paper used in this book complies with the
Permanent Paper Standard issued by the National
Information Standards Organization (Z39.48–1984).

10 9 8 7 6 5 4 3 2 1

660.65
HER
2006

IN MEMORIAM

Be near me when I fade away,
To point the term of human strife,
And on the low dark verge of life
The twilight of eternal day.
Alfred Tennyson, In Memoriam, Stanza L.

For my father
Reuben Herring (1922–1993)

For my brother
Michael McCorvey Herring (1946–1990)

For my twin
Matthew Greene Herring (1952–1953)

For my mother
Dorothy L. Herring (1925–2005)

Contents

List of Illustrations

TABLES

FIGURES

Acknowledgments

At first blush it would seem that the writing of books is a solitary business. We all have images of the misunderstood scribbler, scribbling away at this chosen task. The best quote to capture this is that of William Henry, the Duke of Gloucester. When the indefatigable Gibbon, of *The Decline and Fall of the Roman Empire* fame, presented his second volume to the Duke, William replied, "Always scribble, scribble, scribble! Eh! Mr. Gibbon." So much scribbling did Gibbon do that by the end of the seventh volume, the exhausted author rested his gargantuan hydrocele (what we might call an oversized, severe hernia) on a crutch made for that purpose. The solitary life would seem to have left Gibbon nearly an invalid before he finished his life's work.

Perhaps for Gibbon it was a solitary life. After all, he did not have a laptop, a cell phone, a PDA, or any of a dozen other modern conveniences. Of course, he was also obviously *not* on a deadline. For most contemporary scribblers— or, at least, certainly this one—writing books is not a solitary business. Yes, there are those moments of reflection on a deck in the back yard, in a plane with a laptop, or holed up in a hotel somewhere trying to make sure that, along with the other lives one lives, there is also this one with its own built-in obsolescence (otherwise known as a deadline) that has to be met, and on time. After raising two daughters, I really do not think it takes a village to do so. Having written my seventh book, I can say it takes a veritable army! I could not have done this without that army, and it's only fitting that I say thanks to them now. I quickly add the consuetudinary rejoinder, however, that they are in no way implicated in this book or what it has to say. The views are all my own, not theirs, but I cannot change their names and thank them at the same time.

To Bessie Meeks, my Administrative Specialist, I offer my first, my heartiest, and my most heartfelt thank-you. After almost 30 years. of teaching English to the nation's unwieldy, Ms. Meeks retired some time ago. For reasons no one can understand, she agreed to come to work for me and we have had a four-year love affair over words. Well, love affair is a bit euphemistic. Perhaps "waged war" is more accurate. She curbs, but barely, my words and I argue over all of them. At any rate, thanks so much!

To Douglas Short and Ann Thomas, I extend my heartfelt thanks for the yeoman and woman's work in the Winthrop University's Dacus Library Interlibrary Loan Office. Mr. Short and Ms. Thomas ordered literally scores of books and articles for me, "right now!" as I usually required them. When my fuse blew and my patience wore thin, they managed to get those materials even more quickly and promptly. Of course I understand it isn't easy to tell your boss to take a hike, at least not within his hearing, but they seemed to manage all this and my tantrums with imperturbable calm.

To Dot Barber, my Executive Assistant, and to Larry Mitlin I again offer my thanks for keeping the office running when I was out running about, hither, thither, and yon, chasing various *ignis fatus,* otherwise known as references that could not be found easily or quickly.

To David Weeks, one of the library's resident reference librarians, I offer my condolences. David had the unenviable task of wading through various online sources trying to locate first one thing and then another. David did an extraordinarily good job while working on his own research, as well. Whatever important source I may have missed really isn't his fault.

To my wife of 32 years, Carol Lane, I can only say, "Have mercy on her, Lord." Carol has made my home life so wonderfully blissful that I can spend a few hours every night putting together books like these. I am unworthy of her kindness and devotion but I sure am thankful she chose to dote it on me.

I also extend to my son-in-law, Christopher Slaughter, my heartfelt thanks for his work on the illustrations for this book, all original. Mr. Slaughter may be reached via cslaughter@wsc.ma.edu.

I would be greatly remiss if I did not also mention Winthrop professors Kristi Westover and Dwight Dimaculangan. Kristi and Dwight spent a long lunch with me teaching me DNA 101. I offer my profound thanks. Again, anything I have gotten wrong in this is my fault, not theirs.

Many thanks go to Steven Vetrano, my editor at Greenwood Press, Genevieve Sparacin, my administrative specialist who helped with the index, and to my copy editor Anda Divine, whose painstaking editing has made me sound more literate than I am.

Thanks, too, are proffered to Tony DiGiorgio, President of Winthrop, and Tom Moore, its Vice President of Academic Affairs. President DiGiorgio's wise and steady leadership has made Winthrop the best university of its kind, and has made it possible for people like me to write books like this. Furthermore, it would be hard to think of a better or more supportive vice president than Tom. I have one of the best jobs in academe because of both of these men and their fine and enduring leadership.

Lastly, this book is dedicated to the memory of my father and mother, and two of my brothers. My father spent all of his adult life writing, and it is to his model of the writer's art I owe my own thin talents. My mother passed away somewhat unexpectedly as this book went to press. Like most mothers are to their sons, she was my biggest fan. Her absence will be palpably felt. Michael, a brother six years my senior, passed away in his forties. He was, among much else, a talented writer and artist. Death captured him far too soon. Matthew Greene, my identical twin, is the brother I never knew. He died nine months after his birth. I have felt in some way that I should enjoy life doubly, once for myself and once for him. He deserves to be remembered in some way beyond the solitary marker that designates his small grave in the neighborhood where I grew up.

Preface

When I began this book more than a year ago, genetic engineering, while not a household phrase, still played large in the headlines. Today the phrase has become a household buzzword, if you will, and seldom a day goes by that does not offer some new information, some new strategy, or some new plan to use, manipulate, or otherwise try to incorporate genetic engineering into our daily lives.

We should look on this as nothing but good news, if for no other reason than it moves an important but esoteric topic to the front burner of our intellectual lives. For too many years, recombinant DNA and its offspring have been the bailiwick of a select few. Like hierophants in some mysterious religious rite, practitioners of recombinant DNA and all their ilk have been content to tell the public whatever it needed to know on a need-to-know basis. It has not helped matters that what has been written about genetic engineering has increased almost exponentially in the last 15 years. For example, there have been more books published about genetic engineering in the last 9 years than had been in the 15 before that![1] Even so, no evil intent or even conspiracy loomed here to keep the public in the dark. Rather, the sheer complexity of the subject matter forced those practitioners to report on genetic engineering only when they absolutely had to.

All of that changed when recombinant DNA first changed the food we ate by changing the makeup of that food. It changed yet again when the cloned sheep Dolly appeared on the scene and the once-science-fiction theme of cloning became a near-certain reality. Genetic engineering is now foremost in the minds of many of us, and well it should be. But here's the rub: it still remains a most complicated matter, made more so by the level of education required even to understand its basic parts.

It also does not help that even the protagonists in this debate—the scientists involved in the research—cannot agree on almost anything with the sort of certitude we laypersons require in order to make a decision. Nor does it help that they ascribe nearly everything possible through genetic engineering. Not long from now we will be issuing tee-shirts with the logo, "My Genes Made Me Do It." For example, we recently learned that genes are responsible for the "fickle female orgasm."[2] Genes are also responsible, we have been told, for homelessness, not to mention numerous diseases.[3] Then again, they may not be responsible for all that, and even if we could find a solution for many various gene-dominated diseases, it would not change much because too many human frailties are multifactorial or are caused by much more than changing one gene would cure.[4] For many of us, we are like that city council member (I relate the story in Chapter 9) who, after hearing many hours of debate on both sides, and hearing from an impressive array of experts on both sides, fell back in his chair and said, "What the hell am I supposed to believe?" For every point addressed in this book, there is an equally impressive counterpoint. This not only makes the story hard to tell, but it also makes it most difficult for the reader.

The evenly divided debate teams probably also make it easy for those in the press to tell a story however they wish to tell it, putting whatever spin on it they choose, and relating it as the bottom line or the real story. They are right unless, of course, you take a few minutes and discover that the exact opposite is also the "bottom line" or "the real story." Making any sort of recommendation is, then, most difficult. The few found in this book try to take into account as much of the evidence as possible and then steer toward as much common ground as is feasible.

Having said that, I must add that this book is not a partisan screed. That is, there is no intent here to determine a course of action or side with a particular point of view. Rather, I have endeavored to provide as much infor-mation as possible and then let you, the reader, decide which viewpoint you think is the most fair-minded or the wisest course. In John Stuart Mill's *On Liberty*, Mill provides the perfect excuse for a book like this when he argues that "The only way in which a human being" can know is "by hearing what can be said about it by persons of every variety of opinion" and studying all modes in which it can be looked at. . . . Mill goes on to say that no wise man could do better than this. This book contains every mode and variety of opin-ion I could find. I have included offerings from the best minds who have written intelligibly on this topic, whether first-rate scientists, brilliant journalists, scintillating ethicists, or all three.

Like most such topics, when knowledge and the ability to do a thing out-run our ability to understand the fallout, genetic engineering has outstripped our ability to understand its ethical ramifications. Much effort has been put forth in this book to make ethical considerations more clear. This is only fair. When the Human Genome Project began in the early 1990s, 15 percent of its budget was set aside to delve into ethical matters. While there is no one chapter on ethics, every topic treated in each of the chapters in this book has its own ethical segment.

Because of the aforementioned complexity of the subject matter, most lay-persons acquire their information about genetic engineering from wherever they can. For the vast majority of us, this is the news in its various modalities: newspapers, television, radio newscasts, and electronic postings either via e-mail or the Internet. But many of us, whether we wish to admit it or not, are also influenced by what we see in movies or read in books of fiction. Even though we may view much of this with a jaundiced eye, the fact that much of what was science fiction 25 years ago is now reality has perhaps led us to lend more credence to that which we should not. Consider, for example, the Raelians and their offbeat claims of alien invaders and genetic manipulation.[5] Chapter 1 attempts to capture what we have seen or read over the years with respect to genetic engineering. It will come as no surprise that much of what we have been subjected to has been so blended with truth that separating fiction from it is very nearly impossible. Views in this country compared to views in other countries are about the same with one exception: how we regard agricultural applications of genetic engineering. In other countries, views about genetic manipulation of food are not only more severe but also more radical than our own.

Chapter 2 is the most technical of the nine chapters and contains the most detailed information about recombinant DNA and its scientific process. The chapter begins with a brief history of genetic understanding, from the earliest times and through the more familiar territory of Mendel and his garden peas, a subject taught to most ninth-graders. The discovery of the double helix by the scientists Watson and Crick, however, marks a point in this history where scientific understanding moves ahead of the ability of the general public to keep up. While every effort has been made to render this accessible, aspects of this process are not easily comprehended. I ask for the reader's forbearance.

Chapter 3 picks up the story after the discovery of the double helix. The question became, Now that we know this, what next? The "what next" was recombinant DNA and the promises and heartaches associated with its uses. Some technical information is included here, but far less than in the previous chapter. Stem cell research and the famous Asilomar conferences in the 1970s about the possible downside to recombinant DNA research are also treated.

Are we really what we eat? If we are, Chapter 4 provides some insights into what this may mean for our future with genetic engineering. So-called frankenfoods are discussed and effort is made to try to find a common ground between those who feel any amount of recombinant DNA in food is too much, and those who feel that, regardless of what we add, it cannot possibly do much harm. Differences between Americans and Europeans are also discussed.

Dolly became a household name in the mid-1990s when two scientists, Ian Wilmut and Keith Campbell, cloned her into existence. Chapter 5 unfolds this story and reminds readers that Dolly was only one of many in a long progression of transgenic animals. The chapter also takes up the issue of animal "pharms" and the ethics of using cloned animals for treatment of diseases.

Perhaps every reader of this book is familiar with the Human Genome Project, and Chapter 6 outlines only its highlights. Some effort has been made to provide a historical background for this chapter. To inform readers about the relative genetic complexity, or lack thereof, of our species (we first thought humans were made up of 100,000 genes; it now appears we're only composed of about 25,000—slightly more than the common fruit fly), I relate not only the history but also some of the historical infighting that went on behind the scenes. Looked at today, the Human Genome Project appears to be science working at its best, yet controversies continue to rankle about the project, its funding, and who deserves more of the credit.

Chapter 7 reviews all the medical applications of genetic engineering. Although each chapter contains some medical information, this seventh chapter summarizes all medical breakthroughs and possible treatments in an effort to be as comprehensive as possible about the latest advances in genetic engineering and medical science. The best possible information about the successes and failures are also brought to the foreground, as is the debate about stem cells and embryonic stem cell research.

The short Chapter 8 updates readers on how DNA is now being used in courtrooms across the globe. While every viewer of television's recent spate of detective shows knows about DNA evidence, readers may not be aware of the continuing controversy about it. This chapter compiles the most recent research and provides readers with a map of how to distinguish between DNA evidence and DNA interpretation of that evidence.

Chapter 9 ties together all remaining loose ends, providing readers with the most current information at the time of publication (2006). Here readers will find what new evidence exists about certain medical techniques that were once thought to be fine but now have raised serious questions. Also treated in this chapter is the George W. Bush administration's stance on stem cell research as it was enunciated in the summer of 2005. This chapter contains a survey of the most recent ethical debates at the time of this book's publication. The book closes out with a few recommendations for steering the most common ground possible between the Scylla of partisan proponents of genetic engineering on the one hand, and the Charybdis of partisan opponents on the other.[6] Only readers will be able to tell if, as hoped, I have avoided any hint of bias.

1

"It's Alive!"
Public Perceptions
of Genetic Engineering

No one ever ventured past the graveled road to what the locals called the "turret house" on the jagged crag overlooking the bay. For good reason. Even on a good night it had an eerie quality about it. Tonight it looked especially foreboding amid torrential rains and crackling thunder. The wind blew ferociously. Lightning flashed in frenzied, veined reticulations across the sky. Given its size, you'd think the large house provided haven to some celebrity, but everyone in Darby knew Grayson was no celebrity. Indeed, if anything, Grayson was noted only for his eccentricity. What did he call himself, anyway? An inventor? He had that inventor-scientist look all right, and his shock of Einstein-like hair always looked more electrocuted than combed. And his eyes! Good God, they were beady dots, really, save for the those enormous, horn-rimmed, Coke-bottle lenses he called glasses that magnified them three, maybe five times their actual size. You couldn't look at them (not for long, anyway). Besides, what did he ever invent? Oh, there were the usual kinds of things, like some little mechanized, creature-like contraption that came to the door (assuming the visitor hung around long enough after ringing the bell). Or the conveyor-belt gizmo that retrieved his newspaper and his mail.

Grayson rarely went out, and this only added to his weirdness. If he did go out, he wore a face covering like a male version of a burka. Rumor had it he couldn't go out because of a toxic sun allergy. Most people thought it was because of some terrible invention that went wrong and left his face too disfigured for human company. Grayson never allowed anyone to see him, except for a certain reporter who did a story on him five years before. Word had it that the reporter later took a job in New York but no one ever saw his byline again after that. It made you wonder. Wonder the unthinkable!

Then there were those oddball robots that that were rumored to be all over his house. He may not have allowed anyone to see him, but that did not keep neighborhood children from peering in or the local gossips from enlarging upon what they had seen or heard. Anyway, it was rumored that his oddball robots ran all about his house retrieving things he called for: beakers, test tubes, and Bunsen burners. He called them robots and, to the careful eye, so it seemed. But closer inspection revealed an eerie human quality about them. They didn't speak save for that awful, modulated voice so often associated with computers. Whenever these things spoke, you couldn't help shuddering. The point is, every day at Grayson's seemed like Halloween.

The whole house, indeed, the very air about it, smelled like chemicals— acrid—and it burned the eyes and nose of anyone unfortunate enough to be nearby; well, that is, anyone other than Grayson. Add to all this those godfor-saken sounds that seeped from the sealed windows along with the buzzings, hissings, and terrifying guttural growlings that sent little children running for cover, and you have not a house, but a horror.

For weeks, word had it that Grayson was working on something awful and that he had to be stopped, or at least investigated. That's where I came in. No one in Darby wanted to fool with Grayson. Mayor Standish wanted to send in Dickinson, the sheriff, but the sheriff refused, saying he could not arrest a man for working out of his home. Half the population of Darby worked out of their homes, he reminded the mayor; there was no law against this. Besides, he further reminded the agitated that Tom, who had lost his leg in an awful fishing accident, had gotten back its use through Grayson's robot-ics, or whatever he called them.

Some help. Godalmighty! It looked and acted just like a human leg. It looked too real, functioned too remarkably well. This was no ordinary prosthe-sis. Tom could even run on it, without even a noticeable hobble. Indeed, if you looked at Tom you couldn't tell he had ever been without it. Then came Tom's sudden illness last year. Before the year was out he was dead, barely two years after Grayson had put that thing on. And Tom had no idea how it worked, either, because Grayson had anesthetized him before attaching the awful appendage. Dr. Finehurst said the illness was pancreatic cancer and Tom would have died anyway, with or without the leg. That's what he said, but no autopsy was done because Tom had had to go to Knellenville, where he could get the right treatment. And no one ever saw him again, just like that reporter.

The town hired me to investigate and so there I was, on a night when the thunder shrieked (shrieked, I tell you!) and the wind whipped and howled about that godforsaken crag among trees that never seemed to keep their leaves for more than a month or so each year. And that's another weird thing. That month just happened to coincide with Grayson's annual month-long departure from Darby for what he called a vacation, although no one could tell me where he went or what he did.

I tried the conventional means of seeing Grayson but he would have none of it. Important work, he kept telling me, prevented him from taking even an hour, even a quarter-hour to talk to me. So I took matters into my own hands. And there I was, poking about the brambles around his house—bushes that looked like the hands of skeletons. These damn bushes enveloped his house better than any man-made fortification, I tell you.

With great difficulty I positioned myself by a small window and peered into his workroom, a veritable laboratory of wires and vats and jars—you name it—that lined the walls from top to bottom. How could anyone work in those conditions? They surrounded a large workspace in which a long table stood in the center. I had to wait, but not long, for the lightning to give me a better view. What in God's name was moving on that table? It was something, all right, but what? I couldn't tell from that vantage point because of those scores of jars and the spaghetti-like wires that dangled about the room.

"Mother of God!" I shouted as the lightning flashed in rapid-fire succession. Thunder roared soon after and I stood shaking as two screams rose above it all. Whether one was my own, I cannot say. Although I do not claim special religious inclinations, I can tell you that night I became a praying man. I will never forget that second scream because it literally froze my blood. There in the room stood Grayson, erect, and he bellowed out from the depths of his soul, "It's alive! It's alive! I shall be as God!"

Now this may seem an odd way to begin a book about genetic engineering, but it offers the perfect segue into how we often view medical marvels we cannot begin to comprehend. The above scenario has been played out in scores of movies. Indeed, it is the way we have viewed them for years, even before we began putting them on film. All that is needed is a little over-worked imagination and a heavy reliance on what isn't said, as much as what is. An oddball house inhabited by an overly eccentric figure who keeps to himself doing, well, doing something, but what exactly we cannot say. Throw in very bad weather (always punctuated by lightning flashes and abundant thunder) in a small town inhabited by overly suspicious people, and add a scientist who may or may not be mad, and you have the ingredients of any one of dozens of horror movies of spine-tingling suspense.

But is that really the way we view scientific marvels? Indeed it is, especially concerning genetic engineering. "[F]ictional narratives provide for various types of discourses about cloning and genetic engineering [by] suggest[ing] metaphors, scripts and frameworks that can be used to argue about this scientific advance."[1] In other words, what we glean from popular sentiment often becomes the verbal construct we use in talking, and even debating about, a scientific marvel. In the case of genetic engineering, there is quite a long history of this, especially as it's played out in the cinema.

A brief review of cinematic history will prove the basis of this contention. Whether it's Charles Laughton in the 1933 chiller, *Island of Lost Souls*, Ernest Thesiger in *Bride of Frankenstein*, Humphrey Bogart in *The Return of Dr. X*, or even the ever-hilarious (but in this case, quite mad) Peter Sellers in *Dr. Strangelove, or How I Learned to Stop Worrying and Love the Bomb*, the outcome is always the same: mad scientists and/or physicians mean trouble for everyone else, what with all their infernal meddling.[2] We have innate fears about those who know too much, who know something perhaps they should not know, or know too much for their own (and for our collective) good. It isn't that these individuals are necessarily bad, though many turn out to be. It is that they are forever in pursuit of what Richard Shattuck has called "forbidden knowledge."[3] For all the glorious potential that genetic engineering may possess (and that potential will be fully explicated later), the question remains on the minds of many whether this is the kind of knowledge that mere humans have the ability to comprehend, much less control.

Our popular culture presents us with those worrisome fears and nagging doubts in many ways. For example, few of us over 40 will ever forget the movie *The Blob* (1958).[4] In this B-movie (movies known for their kitsch, poorly constructed and predictable plots, and second-rate—the B-list as opposed to A-list—actors), some kind of knowledge has been released and has gotten out of hand. No one meant for it to, but it did. Efforts to stop it prove futile at first, and many lives are lost before its eventual defeat.

Genetic engineering plays upon that fear because it is at once so complex and so outlandish that the average layman cannot comprehend what it means. An example is the fear that cloning of sheep will lead to human cloning. Can we ever really trust that a small cadre of undoubtedly brilliant scientists can safely and ethically harness such power over our lives?

Mary Wollstonecraft Shelley's familiar *Frankenstein, or the Modern Prometheus* is a good case in point. Although now a universal classic, its 1818 reception was hardly positive. "The foulest toadstool that has yet sprung up from the reeking dunghill of present times," wrote one critic.[5] Of course, others were less disapproving (among them the literary giant Sir Walter Scott) but the story resonated because so many readers felt that science had overstepped it boundaries. Shelley herself may have wondered whether the poor reception was due to her being a woman writer or to the story's subject matter (it was first published anonymously). It didn't help, either, that Shelley's husband, Lord Byron, had already published works reviling religion, or that the couple had harbored George Gordon, whom David Skal calls "one of Europe's most celebrated and reviled libertines." Another factor may have been that Bryon had abandoned his previous young wife for his benefactor's even-younger, 17-year-old daughter Mary Goodwin (Frankenstein's creator), with whom he had eloped. The situation seemed forbidden, irreligious, and vulgar.

The story may or may not have derived from Goethe's Faust, the scientist-scholar who sold his soul to the Devil in order to obtain knowledge no one else had. In any case, Dr. Frankenstein wanted to "ascend to heaven … and exalt my throne above the stars of God," as the Scriptures argue that Lucifer himself desired.[6] That is, he wanted to know what was meant for God, not humans. Dr. Frankenstein stops at nothing, stretches the law and, in many cases, breaks it to achieve his end. He produces a creature, a half-man, half-beast that cannot be classified. Later films tried to humanize the doctor's creation but with little success. In both Shelley's story and the later films, however, the young doctor is monomaniacal about his creation.[7] He is driven to know and to do.

What makes this story so apropos to genetic engineering is how astonishingly accurately it portrays the history of genetic engineering itself. In some ways, only too accurately. In 1998 Jonathan Slack, a biologist at Bath University in the United Kingdom, claimed to have cloned headless frog embryos. This did nothing to dispel the mad scientist image![8] Another early application of gene therapy did not end in a creature per se, but it did end in death. The brilliant young scientist responsible for the feat not only broke laws but also even equivocated about his research in an effort to know, to be successful. So arrogant were his efforts that he was later barred from the field.

Another very famous story springs to mind, this one from the pen of the well-known and highly famous Robert Louis Stevenson, *The Strange Case of Dr. Jekyll and Mr. Hyde*. In this case, the scientist inflicts the pain on himself rather than on others by drinking a strange and unknown potion. It overtakes his mind and even alters his physical appearance. However, the outcome endangers everyone in his path, especially young women, as the half-human Hyde pursues them relentlessly. On the other hand, the debonair Dr. Jekyll is not only well-liked but also well-received in society. Yet when he drinks his strange concoction (or is it in the genes?!), he is no longer himself. He is so altered, even in appearance, that he becomes possessed by his discovery and cannot control his actions.

Stevenson wrote in metaphor, allegorizing his story to stand for something else. But the story would have very little appeal if, for example, his main character had been a postman or a dogcatcher. It is because he is a scientist that the story works. The human mind is automatically and credulously drawn to the scientific profession that itself seeks to know the unknowable and think the unthinkable. As we will see later, these same sentiments have been leveled at genetic engineering by its critics. Proponents of this strange new science cannot predict outcomes or provide full assurances that what is discovered can or will be controlled to the ultimate satisfaction of the public it seeks to persuade or reassure.

Stevenson patterned his story on the true story of Deacon Beodie, an Edinburgh prelate who maintained his respectability by day but became a terrifying murderer by night.[9] Such history does not reassure; Stevenson changed the profession to make the story more plausible. Dr. Jekyll's charming persona is chemically altered so that his alter idem, Edward Hyde, becomes, as Stevenson describes him. " half-human and troglodytic" in his madness. Hyde shrinks from his human status to one that Stevenson takes pains not to be too specific about: "He is not easy to describe. There is something wrong with his appearance; something displeasing, something downright detestable. I never saw a man I so dislike, and yet I scarce know why."[10] The unknown or the unknowable is made all the more frightening by this nondescript characterization. We see the same sort of technique in *The Cabinet of Dr. Caligari*, where a hodman hypnotist attempts to control his murder-and-mayhem sleepwalker.[11]

Jewish literature is replete with similar stories, such as *The Golem* (1921). Here, another scientist-like meddler creates a piecemeal, man-like monster. The story of the golem has frightened children for decades. Likewise, the 1916 six-hour serial film *Homunculus*[12] features an artificial superman who, upon learning his origins, becomes unchained and goes on a murderous and destructive crusade. Then there is the 1922 film *The Monster*, which depicts Dr. Ziska, a monomaniacal surgeon who lures unsuspecting motorists near his sequestered home on Long Island.[13] The doctor evokes all of our fears with his hidden doors, private rooms, and a cold, sterile operating table. There is even an electric chair! This review would not be complete, however, without the appearance of the terrifying Lon Chaney in *A Blind Bargain* (1922). Chaney plays the roles of both surgeon and his subhuman, hunchbacked (naturally) creation.

On the other hand, physicians representing science are not always relentlessly portrayed as evil. For example, Somerset Mangham's *Of Human Bondage* (1934) depicts the physician as doing good.[14] Another similar movie glorifying physicians is *The Country Doctor* (1934),[15] where physicians are regarded as being just short of angels. Other movies, such as *The Story of Louis Pasteur* (1936) and *Madame Curie* (1943), continue the short-lived trend of physicians and scientists being viewed as altruists at worst, and great healers at best.[16] Indeed, films throughout the 1930s and 1940s portrayed physicians as altruistic, while their brother scientists did not always fare so well. But this did not last very long for physicians, especially when they began venturing into the unknown.

By the 1950s, popular sentiment as evidenced by cinematography released a series of mad-scientist films involving living brains in giant petri dishes, or living somethings in bell jars.[17] From these films emerged a distinct, cavalier nonchalance on the part of scientists themselves about what they do. They are "doing science" and, for that reason (as guardians of the new Grail) cannot possibly be doubted, challenged, or curtailed in any way.

But we need not dwell on the distant past. Sentiment regarding scientists and physicians has hardly changed at all in the last three decades. Whether we view the 1958 film *The Fly* or its more recent (1986) remake starring Jeff Goldblum, the outcome is the same: mad, obsessive scientist with teleporter Cuisnarts his molecules into a half-man, half-fly creature. Even as he begins the awful metamorphosis that rivals Kafka's cockroach, Andre DeLambre, the lunatic scientist, cannot be persuaded to stop. The power of knowledge is an aphrodisiac unlike any other. Even beautiful, eager-to-please young women cannot drag him away from his machines. His experiment will succeed, as if success is the only thing that matters.

Gross Anatomy (1989) depicts self-absorbed medical students and *Flatliners* (1990) features physicians who toy with death. Fortunately, in the latter case it is only their own deaths but, even so, one cannot watch the movie without thinking that these young physicians are deranged by science, mad with self-adulation, and intoxicated with arrogance.[18] By 1997, physicians are *Playing God* in a dreadful movie that showcases gallons of blood and stupid dialog.[19] Even so, it underscores popular sentiment about the heights to which scientists and physicians will aspire. The theme of scientists who attempt to play God will be repeated by critics of genetic engineering.

Finally, who can ever forget Steven Spielberg's *Jurassic Park*? The movie not only focuses on genetic engineering, but also revolves around advances made with recombinant DNA to achieve new ends. Equally unforgettable, however, is the horror with which the movie ends, as the arrogant scientist cannot control the monsters he made or close the lid on the Pandora's box he opened.

But is this fair? Is it right to begin a book about genetic engineering with the idea that some group (scientists in this case) have arrogated for themselves the roles of judge, jury, executioner, or implementer? The Asilomar conference of 1975 (further discussed in Chapter 3) found this to be the very horns of the dilemma. David Baltimore opined, "We're stuck between self-determination of limits and the imposition of orthodoxy. We're stuck between the self-interests of scientists and the public interest."[20] Only scientists can determine whether the science is right. Only scientists can determine, so the argument goes, how far to go or what risks should be taken. And only scientists are capable of implementing genetic engineering, but are they the right ones to determine its overall public good?

While it may not be "Damn the torpedoes, full speed ahead," there has been a sense of "Pay no attention to the man behind the curtain," as the wizard said to Dorothy in *The Wizard of Oz*. The general public has been asked to ignore certain somewhat frightening potential scenarios, along with outright failures. We have been asked to forget those unfortunate events and focus on the overall goal. It may not be "The ends will justify the means," but it sounds a lot like it.

Of course, while the general public cannot possibly be asked to understand the incredible complexities of genetic engineering, "[I]t is [also not] possible to exclude [it] in a matter that should be of public concern."[21] Failure to understand should not vitiate the public's right to voice its concerns and be heard. I have begun with an overview of popular culture for two reasons. The first is to describe where public sentiment is on this issue. The second is to delineate for scientists just how high is the hill that they must climb in order to gain both public acceptance and public trust.

In 2004, California voters passed a pro-embryonic stem cell research measure. For many commentators, this proved to be a major bit of news. But what the vote really means is that the public is willing to see such research pursued. However, the vote says nothing about public sentiment following a colossal failure.

The strength of current public sentiment about genetic engineering is rooted in how popular culture views, and has viewed, scientists. If there were but a handful of movies evenly divided between good and bad scientists, another argument would be required here. But an overwhelmingly larger number of movies portray scientists as very nearly irredeemably bad. Scientists, at least in their chosen persona of Pandora or Eve or Prometheus—that fabled demigod who sought to bring fire (representative of knowing too much)—are portrayed almost exclusively in a bad light.

Perhaps popular culture is wrong. Can we be sure that the view of popular culture is not merely one of many of Hollywood's mythologies? Certainly Hollywood's penchant for stretching the truth or fabricating cannot be discounted. Certain aspects of genetic engineering, however, have been roundly rejected. For example, a recent Market & Opinion Research International (MORI) poll found that 77 percent of those questioned in the United Kingdom are skeptical about genetic engineering and would like to eliminate genetically modified organisms (GMOs).[22] Even among their own number, scientists have pointed out that the long- and short-term effects of biotechnology are inherently unpredictable.[23] In the United States, it has been estimated that nearly 80 percent of Americans would not eat genetically modified foods if they were labeled. The fact that they are not is a matter we will examine in more detail in Chapter 4. Indeed, there has been an increasing skepticism—no, make that rejection—by the public of the use of this technology, at least with respect to food, over the last 10 years.[24] Part of this has to do with the idea of perfecting or improving the human race. As Kurt Batertz put it, "Hardly anybody … who reflects upon [the betterment of the human race] will be able to do so without sensing a certain uneasiness."[25] In fact, it is this notion of betterment that is at the heart of the genetic engineering controversy.

But doesn't this run counter to the recent vote in which Californians overwhelmingly approved continuing stem cell research? Was that merely the exuberance of cisatlantic inhabitants?[26] Actually, it was not. First, California has long been known for its willingness to accept the unconventional and nontraditional and is thus hardly representative of the nation at large. Second, the United States recently re-elected a president by the single largest vote of any president ever, who remains foursquare against an "anything goes" open research effort in genetic engineering. Third, the American people as a group have never rejected on face value more research into anything. Finally, no one really believes, as vice-presidential candidate John Edwards claimed, that any sort of genetic engineering research will enable any wheelchair-bound individual to get up and walk, at least not in our lifetimes. Even those who think that Americans and Europeans are divided on the issue of genetic engineering (with the Yanks being only slightly more progressive) agree that while public sentiment "is the ultimate driving force behind science and technology funding," it is also the driving animus "behind regulation, political opposition and drawn-out court battles" in both countries.[27]

A 1998 Wellcome Trust Report examined the power of themes and motifs in books and movies that have aided and abetted a view, often a negative one, of genetic engineering. Although it was not always clear which themes and motifs respondents to the Wellcome Group were trying to make with respect to genetic engineering, they chose them in a manner to which "others within the group would relate."[28]

Recent Gallup polls may help shed more light on what seems to be an on-again, off-again love affair with genetic engineering, not only between the United States and Great Britain, but within our own country. In early polls, the U.S. Office of Technology Assessment (OTA), a nonpartisan group of experts who have been charged with helping Congress unravel the myriad of opinions about genetic engineering, found widespread agreement and dissent. The OTA reported that "The survey finds that while the public expresses concern about genetic engineering *in the abstract* [italics mine], it approves nearly every specific environmental or therapeutic application. And while Americans find the end products of biotechnology attractive, they are sufficiently concerned about potential risks, that the majority believes strict regulation is necessary."[29] As long as Americans can actually see various applications and the hoped-for outcomes, they are fine with the idea. But in the abstract, Americans are fearful that genetic engineering may have deleterious consequences and so it must be regulated, either by government, a special task force of scientists, or a combination of both.

What should be of more concern, however, is the amount of knowledge (or, rather, lack of it) the public has about genetic engineering in general. In a 2002 Roper poll, only 23 percent of American adults said they knew "a great deal about genetic engineering," while 57 percent said they knew "something but not much," and 20 percent said they knew "nothing at all."[30] When asked if it was possible to use genetic engineering to change a baby's genetic makeup before it is born to prevent it from having a genetic disease, a full 78 percent (more than two-thirds) of respondents said either no, it could not be done, or they did not know at all.[31] Nevertheless, 66 percent of Americans in that poll were still confident that they know the meaning of the phrase "genetic engineering."[32]

Although not many respondents knew whether the government did regulate genetic engineering (70 percent said it did not or they did not know if it did), 71 percent felt that government should be regulating "the quality of genetic engineering." The nation is divided about whether this regulation should be more strict than it is (37 percent) or is about right (33 percent). Still, 27 percent (nearly one-third) of the respondents either think government should stay out of it altogether or they aren't exactly sure what those regulations should be. When it comes to the issue of cloning, however, Americans are hardly divided. When asked if the government should limit the cloning of humans, 84 percent said yes. Only 11 percent said the government should not.[33]

Interestingly, when asked in this same 2002 survey about why topics in genetics are worrisome (techniques such as in vitro fertilization, cloning, genetic testing, and genetic engineering), more than one-third of respondents said using these technologies "is too much like playing God."[34] Another one-third (35 percent) said that these new technologies "can easily be used for the wrong purposes." In other words, almost 70 percent, without actually saying so, thought scientists had opened a Pandora's box and unleashed knowledge that either should have been forbidden or was knowledge that should not be made too widely available for mankind. Most respondents (41 percent) were persuaded by the prospect that certain genetic diseases "can be wiped out," and so they hope for continued research and testing. Moreover, when asked how they couch the issue of genetic engineering, 54 percent tend to do so in a health and/or safety context. Still, one-third of respondents at that time thought of the issue of genetic engineering in moral or religious terms.

While Americans would be happy to see disease wiped out via genetic engineering, 40 percent think tinkering with food, animals, fish, trees, or plants creates too many risks to continue. Sixty-four percent think that genetic modification will contaminate ordinary plants and is too risky to continue, while 57 percent fear it will create too many "superweeds."[35] In the end, however, Americans are overwhelmingly in favor (84 percent) of continuing genetic engineering. [36]

What are we to make of all this? Even a studied reading of these results yields very few conclusions other than that genetic engineering should continue, but very slowly. While these recitations of data may have been wearying for some readers, they reveal one clear, if startling, picture: Americans are ready to proceed but only under careful, studied, and regulated conditions. What makes this debate so interesting is that, unlike so many other issues—abortion, for example—where there are clear sides, genetic engineering proffers none. Rather, the "side" that genetic engineering offers to scientists working in the field is: make a mistake, take too many liberties, or act without regard to human life, and the sentiment of the American public will be fiercely opposed to you. On the other hand, proceed with caution, be careful of the preciousness of human life, and never go forward hurriedly, and the sentiment of the American public will remain positive—if not arm-in-arm, at least hand-in-hand. If this sounds to you like Americans are of two minds on the topic, you are right.[37]

In many ways genetic engineering is the proverbial landmine of dissent: benign on the surface, but one false step and KABOOM! The connection between popular culture and scientific discovery comes to this: "It took communication scientists and other social decision makers many years to learn that objectivity in the world of media and a computable measure for handling risks in public communication stem from risk. Public risks are mainly constructed by the media, by their selection criteria and newsmaking routines."[38] To put it another way, the manner in which this debate is talked about is often the product of what is seen in the media, whether it be in movies, books, or the newspapers. It will be important to remember this as each new chapter is considered here. Public sentiment must struggle with its altruistic sentiments to help others, on the one hand, while battling against visions of headless frogs and half-human clones cast aside like so many cars in a junkyard of spare human parts, on the other.

2

History of Genetic Engineering from Mendel to (Genome) Maps

We begin this chapter with three familiar myths. Myths have long been misunderstood by the general public. Some confuse a myth with a falsehood but this isn't the way the ancients understood them. Rather, a myth was a story that enlarged upon truth or served as the vehicle for it.[1] The three we will look at, Icarus, Pandora, and the fire in the lap of God, illuminate a truth about genetic engineering that will be important to remember as we proceed.

Ovid's tale of Icarus is familiar. His father, Daedalus, known for his most fecund creative efforts, longed to get home; he was growing quite tired of Crete where he was stranded as King Minos's inventive genius.[2] Exasperated, he told the king who held him that while he blocked the land and ocean, he could not block the sky. Or as Daedalus put it, "Minos may possess all the rest, but he does not possess the air." Daedalus then set his mind to sciences never explored before, and altered the laws of nature.[3] He collected feathers and began to fashion them into wings for himself and his son, Icarus, to escape

At first, all seemed to go well as Daedalus balanced his body and hovered in midair. He prepared his son to fly with a fulsome warning, "Remain midway between Earth and heaven." Too far toward the sun, and its heat would scorch the feathers and melt the wax that bound them. Too low, and water from the oceans and ponds would make them too heavy. His final words were, "Follow me!"

Away father and son flew, homeward-bound at last. Onlookers must have thought them gods as they flew midway between heaven and Earth. The father tried to keep Icarus close but soon the son found the elation of flying too great to bear. His elation was short-lived. As he flew into the open sky, he soared ever-upwards until he flew too close to the blazing sun. The wax melted, the feathers fell like snow, and Icarus plunged to his untimely death. When his father saw the feathers scattered about the ocean's surface, he cursed his inventive skill.

Pandora (from the Greek, meaning gifts) may be one of the most familiar myths of all for it has been retold in nearly every elementary school textbook. Zeus had given Pandora to humankind, its first woman, as a punishment to Prometheus for stealing the gods' fire. Her many gifts made her irresistibly alluring. Pandora possessed, in addition to her exquisite beauty, an insatiable curiosity. She simply had to know everything. This led her to open a most beautiful bejeweled box, though she had been warned not to.[4] One day she lifted the lid and out flew all the ills of the world: hatred, greed, death, diseases, envy, and so on. Upon seeing her mistake, her despair proved too much and she hurriedly closed the lid, trapping only Hope, enough perhaps for herself and all the world.

Our final myth is perhaps the least familiar of the three, possibly because its origin is non-Western. But of the three, it may have the most to tell us about ourselves with respect to genetic engineering. It comes to us through Hebrew literature and what is known as the Kabala (or Cabbala), a rich source of Jewish myth. It seems that one day God was sitting about after five arduous days of creation, thinking what to do next. He looked at the fire in his lap and began to ask it for counsel. The conversation might have gone like this:

"I have made my crowning achievement—the creation of man, made in my image, just a little lower than the angels," exulted God.

"So, in other words," said the fire, "he is able to think and to reason?"

"Precisely!" said God with some gusto.

"This is not a good idea," said the fire.

What can these three myths possibly have to do with genetic engineering? They illustrate for us the controversy that lies at the bottom of this debate. The death of Icarus illustrates mankind's quest to know the impossible, as well as an arrogance regarding his ability to know, even if it should put him or others in danger. Pandora's box represents mankind's insatiable curiosity regardless of the outcome, be it good, evil, or a mix of both. Finally, in a humorous fashion, the fire in the lap of God represents God's mercy on the one hand, and our own self-destructive tendencies on the other. Perhaps we humans weren't such a good idea after all, given our avid desire for knowledge of that which helps, as much as that which hurts.

A wag once said that as information doubles, knowledge halves while wisdom quarters. Genetic engineering may well be the great knowledge but it requires a rare wisdom we have yet to learn. It also possesses tremendous power for good or an equal power to harm. The discovery of a simple genetic marker may one day reverse cystic fibrosis or horrific death by Lesch-Nyhan Syndrome.[5] It can also unleash a biochemical gene that could kill more people than all our wars combined in a matter of a few days. Our behavior so far leaves the proposition unclear as to outcome. So, how did we get here, on this the horns of our genetic dilemma?

A BRIEF HISTORY OF GENETICS[6]

Although the history of genetic engineering is only 100 years old, humans have been tinkering with genes for thousands. Most readers will remember Gregor Mendel from their ninth-grade biology, the fairly obscure Austrian monk who discovered heredity by using simple green peas. More about him later. But the history of genes (heredity, really) was discovered much earlier than Mendel, by about 5,000 years.

The Talmud, the great code of Hebrew law, has a very curious injunction that runs something like this: do not circumcise a male whose maternal uncle was a "bleeder," for he could bleed, too.[7] They understood that descendents from such males would likely carry what we have come to know as hemophilia. To circumcise such a male would lead to his sure death. But there is more. A passage in Genesis depicts Jacob and Laban divvying up land and turns on Jacob's rudimentary knowledge of how genetics—though, of course, he would not have called it this—worked in common farm animals.[8] It's unclear from the text whether Jacob knew this intimately enough to credit him with anything more than dumb luck. Even so, the story remains as a testament to a working understanding of heredity in a very early context. Circa 1000 B.C., the Babylonians celebrated the pollination of date trees with religious dances and such, while the ubiquitous Aristotle (323 B.C.) speculated on the nature of reproduction and inheritance.[9]

Later, Shakespeare in his play *Titus Andronicus* reveals his own, again fairly rudimentary, understanding of heredity when he writes:

Peace, tawny slave, half me and half thy dam!
Did not thy hue bewray whose brat thou art,
Had nature lent thee but thy mother's look,
Villain, thou mightst have been an emperor:
But where the bull and cow are milk-white,
They never do beget a coal-black calf.[10]

Shakespeare displays what observation has taught him, or at least what someone around farm animals taught him. The English physicist Robert Hooke wrote an uncanny book, *Micrographia,* which contained more than 50 plates of his microscopic examination of blood droplets that revealed more than he expected.[11]

While all of this is very interesting, our main concern comes much closer to our own time. The word "gene" did not appear in lexical use until 1913 in Dorland's *Medical Dictionary.*[12] The word itself was coined by a Dutch botanist, Wilhelm Johanssen, in 1909.[13] The term that is the subject of this book, genetics, appeared only 15 years earlier, in L. F. Ward's *Outlines of Sociology.* Of course, these words, derived as they are from the Greek, have close relatives in "genesis" and "genealogy." Obviously, all the terms have significance in reference to our origins in one way or another.

The scientific beginning of this study, however (the Talmud notwithstanding) comes to us from that somewhat obscure Austria monk mentioned earlier, Gregor Johann Mendel (1822–1884).[14] While we must not omit the Dutch genius Anthony van Leeuwenhoek's (1632–1723) invention of the microscope in this connection, and thereby the discovery of the entire microscopic world (see Hooke, above), it is Mendel who begins our story.[15]

In an 1865 meeting of the Natural Sciences Society of Brno, Mendel presented the first account of his decade-long study of plant hybridization. His paper, one of those singular moments in science, made its way into the Society's publication a year later but went completely unnoticed. Three and a half decades later, three other scientists (Hugo de Vries, Carl Correns, and Erich von Tschermak), working in the same area but in three different cities, published within months of each other the same results Mendel had already discovered. None of them knew they had been scooped by a monk who lived in relative obscurity. Only then did Mendel, whose research came about the old-fashioned way via excruciatingly hard work and near-constant observation, become the so-called father of genetics.

From 1856 through 1863, Mendel conducted experiments on plants, not just on some but as many as 30,000! He collected, sorted, observed, labeled, and counted, giving rise to the idea held by many that Mendel had a fairly sophisticated inductive hypothesis in mind that he tested and retested. He was, after all, an excellent mathematician and in letters to fellow scientists referred to his experiments as "statistical relations."[16] He developed seven plant characteristics from all this work and proposed the idea that heredity is particulate, or comes about through a series of separate units or pairs, rather than blending inheritance, what was then thought to be the case.

Particulate heredity gave way to now-familiar characteristics discovered by Mendel, dominant and recessive traits. Mendel called them elements, later to be called genes.[17] Sadly, even after Mendel's publication in the Society's pages, no one noticed for nearly half a century. It would not be the first time that

Gregor Johann Mendel. From the National Library of Medicine, History of Medicine Collection.

genetics or related studies would be discovered long before the scientific community noticed. To be fair, it is likely that none of Mendel's fellow scientists understood what he was saying. Of those who did, it so contradicted so-called received wisdom that it could not be believed.

In 2003, Lee Keekok said that although "Molecular genetics is said to be at a more basic, and therefore, deeper level of understanding" of genetic material than [Mendel's Laws], what he left us cannot, should not be minimized."[18] Heredity came in pairs and offspring received one from each parent. The pairs came in dominant and recessive types. Mendel's discovery was paradigm-changing. Mendel showed how traits are passed from parents to children through gene types. He even explained mutations. While we are quick to think of mutations as always incorrect or wrong, we'll see that this isn't always the case.[19] Mendel's most famous legacy is summarized in Table 2.1 below, and will be familiar to most readers.[20] In this case, the Mendelian laws of heredity show that the dominant brown-eyed (B) gene (or in Mendel's case, element or particulate) occurs more frequently than the recessive blue-eyed gene (b). Both the BB pairing and the two Bb pairings yield offspring with brown eyes. There is only a 1 in 4 chance that from one pairing there will be an offspring with blue eyes. This becomes important because humans desire to know why certain things occur, whether they attribute it to science, old wives' tales, or social ritual.[21] Mendel's so-called laws provided a perfect explanation while also opening the door to further study. Mendel provided an explanation about why certain characteristics breed true (in genetic parlance) and why some recessive traits eventually show up, or why all might appear in later generations but not necessarily in succeeding ones.[22] It was truly a remarkable feat.

In many ways, Mendel originated gene theory, or what came to be called the theory of transmissible factors.[23,24] This led to our understanding of how cells (revealed by Leeuwenhoek) contain copies of each gene, how fertilized cells (called zygotes) contain two alleles or variants of genes, and how the genetic makeup of an organism (the genotype) is different from its actual traits or characteristics (its phenotype).[25] Even granting all this, we must not credit Mendel with too much.[26] Indeed, there is some controversy about his work, including that his data are too good to be true and that his experiments are, at the very least, fictitious or questionable.[27]

Mendel's very specific work with pea plants yielded this knowledge because pea plants contained a predictability that other plants did not. Moreover, they were readily available, hardy annuals that were easy to grow, that insects would cross-pollinate, and were already established with obvious differences (e.g., tall, dwarf, white or red flowers, seed either round or wrinkled, green or yellow).[28] It helped enormously that Mendel's work appeared very nearly contemporaneously with Darwin's *Origin of the Species*. In fact, it would appear that Darwin (along with 133 scientific associations Mendel sent it to) got a copy of Mendel's paper but did not even bother to read it.[29] While the two ideas did not merge at once (Darwinians wanted Darwin's theory of evolution to explain everything), it did not take too long for the two ideas to be seen more as hand-in-glove than as separate, competing scientific theories.[30] Indeed, much later Dar-

Table 2.1
Mendel's Table of Laws

Parent	Bb	Bb
Bb (eye color)	BB	Bb
Bb	Bb	Bb

win himself wrote in a letter to the English botanist Sir Joseph Dalton Hooker, stating, "I have read heaps of agricultural books, and have never ceased collecting facts. At least, gleams of light have come, and I am also convinced (quite contrary to the opinion I started with) that species are not (it is like confessing to a murder) immutable."[31]

Think back to Leeuwenhoek's microscope and his subsequent opening of the world of previously unseen life. As microscopes improved, scientists got a much better idea of cell structure and cells' component parts: mitochondria, vacuole, rough and smooth endoplasmic reticula, lysomes, and Golgi complex, as seen in Figure 2.1 below.

Our attention, however, must focus on that most familiar entity, the nucleus. Inside the nucleus, along with its two-unit, fluid-filled spaces, lies the chromatin, or that portion of the cell nucleus that contains the DNA. Inside are familiar-looking entities that remind us of Xs or Ys, the pairing of which determines gender. Most cells in our body have 46 chromosomes, 23 from our mothers and 23 from our fathers. Let's look at one of these, a chromosome. It is the chromosome that contains enormously long (see Figure 2.2), spaghetti-like strands of deoxyribonucleic acid (DNA). DNA controls cell metabolism and all of our heredity. DNA has the two quite familiar long strands of nucleotides, twisted together and containing enormous amounts of information, called nucleotide bases, or simply bases. Each base is designated

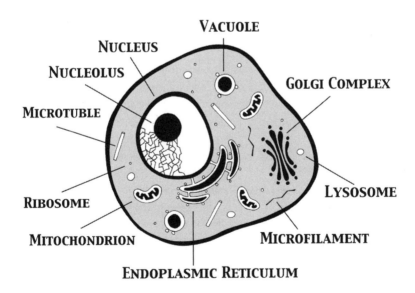

Figure 2.1   Major cell structures

SIMMS LIBRARY
ALBUQUERQUE ACADEMY

by the letters A, G, C, T, which will be further discussed below. The amount of information on one chromosome set is about six billion units of genetic code, an average of 130 million for each chromosome. To get some idea of how big (and small) all this is, if one cell is stretched out it contains about three yards of DNA, all compressed in an area that is two-millionths the volume of a pinhead.[32] The period at the end of this sentence is larger than the space occupied by DNA.

In close proximity lies the nucleolus that contains the other single-most important molecule in our study, ribonucleic acid (RNA, often written mRNA for "messenger RNA"). RNA contains the codes or the blueprints that DNA uses for its orders, referred to as transcription and translation. As said above, DNA is in every cell in the body except red blood cells. While genes (about 20,000 to 25,000 different ones in humans) are most commonly associated with chromosomes, genes are found elsewhere, namely the episomes and plasmids in bacteria, which are small, usually circular pieces of DNA.[33] Plant cells have been found to have their own genes, inherited independently from chromosomes.[34] It's all marvelously complex and yet singularly simple. So how did we come to know all this and, furthermore, to know how all this works?

The simplest answer would be to cite Watson and Crick's now-famous 1953 work on the double helix (about which more will be said below), and that would be right. But that leaves out too many other helping hands and heads along the way. For example, there is the Scandinavian scientist Willard Johannsen, asked by his colleague geneticists in the 1920s to refrain from referring to Mendel's particulates and to begin calling them genes, instead.[35] Niels Bohr (1885–1962), the Danish nuclear physicist, theorized in the early 1930s that we could not account for all biological phenomena in the same way as we had before. There were, he argued, physical and chemical explanations that defied conventional wisdom. In the mid-1930s a student of his, Max Delbruck (1906–1981), published a paper that called this unconventional wisdom the science of genetics.[36] Johann Friedrich Miescher, a Swiss researcher, was the first to point to the special properties of nucleic acid and a chemical unlike any other, one he called nuclein.[37]

However, it was the work of James Watson and Francis Crick that unified the approach to genetics and made DNA a household word, leading to what would become the whole genetic engineering industry we now face today with either great enthusiasm or great hand-wringing. More than half a century ago, before computers, Palm Pilots, and fax machines, Watson and Crick published their grand treatise in the highly prestigious *Nature* magazine in 1953. Crick had been trained as a physicist and Watson trained as a phage geneticist. Like Delbruck, Watson held no brief for classical geneticists, whom he dismissed as "nibbling around the edges of heredity rather than trying for the bull's eye."[38] Watson and Crick's work has been considered a discovery as important as Mendel's or even Darwin's. This has been confirmed by their sharing of the Nobel Prize in Medicine/Physiology in 1962. Watson was a mere 25 years old and a postdoctoral student at Cambridge when he and Crick wrote the hallmark paper.

Watson and Crick were outsiders to biochemistry.[39] Watson had seen x-ray patterns of DNA produced by chemist Rosalind Franklin and physicist Maurice Wilkins (King's College, London), after Franklin had lined them up in a thin glass tube and forced x-rays through them.[40] As soon as he saw them, Watson knew he had seen something extraordinary, saying later, "The instant I saw [Franklin's x-ray] picture, my mouth fell open and my pulse began to race."[41] What caused the uproar were the two chain-like sequences that formed a twisting staircase, or the so-called double helix. The two strands (sugars, known as deoxyribose and a phosphate molecule) formed the two sides of the staircase or ladder. This is often referred to as the phosphoribose backbone of DNA. The rungs of this ladder are formed by nucleotide bases mentioned above. The Watson and Crick model did much to set the stage for the next level of developments. It not only explained what took place but also provided a built-in explanatory mechanism for how heredity works. The very nature of the model suggested to Watson and Crick that this DNA had another component that may well hold the key to life; indeed, Crick said as much when he walked into the Eagle Pub in Cambridge and announced, much to Watson's chagrin, "We have found the secret of life!"[42] In their astonishingly short *Nature* piece, Watson and Crick point out that they were not unaware of the helix's copying capabilities.[43]

The Watson-Crick model, as it came to be known, is stark in its simplicity, especially given all that DNA is responsible for bringing to pass.[44] Each molecule contains simple units known as nucleotides. These are composed of materials that have been known to those studying genetics for years. As indicated above, there is a so-called sugar arm (sugar deoxyribose) made up of several oxygen atoms congregated about a phosphorus atom. Four other elements make up the other rungs of the ladder: adenine, guanine, cytosine, and thymine. These are now generally abbreviated as A, G, C, and T. The double helix in Figure 2.2 shows this arrangement.

Just finding these four elements proved to be a monumental feat in the development of disease-conquering medicine. Each base denoted by these code letters is 1/50,000,000 of an inch in diameter![45] But breaking this code and discovering what it all meant was a Herculean effort. The letters explain how protein is generated in the body. Much like flour, water, and yeast combine to

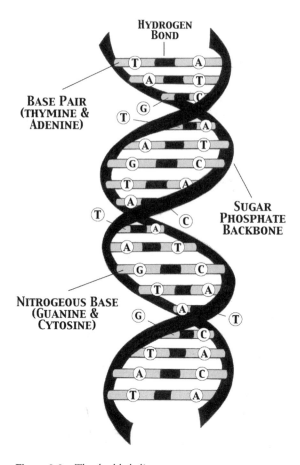

HYDROGEN
BOND

BASE PAIR
(THYMINE &
ADENINE)

SUGAR
PHOSPHATE
BACKBONE

NITROGEOUS BASE
(GUANINE &
CYTOSINE)

Figure 2.2   The double helix

make dough, these letters combine to make proteins (this is the way the body rebuilds its building blocks). Three of these element rungs, often referred to as triplets, together form a genetic code, or codon (the triplets), for an amino acid or a stop message. Taking four of these codes, three at a time, yields the 64 possible protein combinations.

The four bases not only protrude from what could be called the handrails, but also stick to each other or, rather, to their complementary pairs. After reviewing the Franklin data (about which Watson was later not very complementary), the two men realized that the chains or handrails could line up in opposite parallel fashion.[46] They also discovered something even more important about the bases. Adenine (A) always paired (or stuck) with thymine (T), while cystoine (C) always paired (or stuck) with guanine (G). Like salt and pepper or black and white, the A–T and G–C complementary pairs are nearly always predictable (exceptions are mentioned below). The shape or structure—

Dr. James Watson speaking at the NIH Conference on DNA December 15–16, 1977. From the National Library of Medicine, History of Medicine Collection.

the helix—is beautiful not only because it is so simple but also because, as indicated above, it provides a ready-made explanation of how the genetic code is carried and copied. If it were not for this process, no material could be transmitted from one cell to another, much less from parent to child.

These complementary pairs also allow us to make certain predictions. If, say, we find a strand (referred to as a sequence) that reads ACGTCTCTATA, then we know what must be joined to it in parallel fashion is a strand (or sequence) that reads TGCAGAGATAT, if the first is supraimposed above the second. This also led geneticists to refer to the code as palindromic, meaning "the same backwards as forwards" as, for example, the word radar. In the genetic world, this means that one parallel side of the rungs on the handrail predicts what its opposite will look like (while restriction enzyme recognition sequences are palindromic, not all DNA is). Codons are also palindromic because when they are read one way they stand for one acid, but when read backwards they stand for another. Table 2.2 below shows how the same triplet can be one acid or, read differently, another. What this means is that DNA contains not so much a mechanism for copying itself as it does for copying its opposite!

But how does this work? When a copy or copies of DNA are needed—as, for example, when you cut yourself and the skin must heal—20 active enzymes are called to order. These enzymes tell the paired bases (A–T and G–C) to cease their unity.[47] At once, the bases let go of each other and begin what is known as an unzipping process so that the double helix has exposed both sides of the single strands or handrails with half-rungs (that is, one side will have an unattached A the other an unattached T; one with an unattached G, the other an unattached C, and so on). The two new strands (sequences) with their exposed As, Ts, Gs, and Cs will then match back up, forming a new strand (in this case, new skin) all over again.

Table 2.2
Amino Acids

Triplet codon (DNA)	Amino Acid
GCT, GCC, GCA, GCG	Alanine
CGT, CGC, CGA, CGG, AGA, AGG	Arginine
AAT, AAC	Asparagine
GAT, GAC	Aspartic acid
TGT, TGC	Cysteine
GAA, GAG	Glutamic acid
CAA, CAG	Glutamine
GGT, GGC, GGA, GGG	Glycine
CAT, CAC	Histidine
ATT, ATC, ATA	Isoleucine
TTA, TTG, CTT, CTC, CTA, CTG	Leucine
AAA, AAG	Lysine
ATG	Methionine
TTT, TTC	Phenylalanine
CCT, CCC, CCA, CCG	Proline
TCT, TCC, TCA, TCG, AGT, AGC	Serine
ACT, ACC, ACA, ACG	Threonine
TGG	Tryptophan
TAT, TAC	Tyrosine
GTT, GTC, GTA, GTG	Valine
TAA, TAG, TGA	Stop codons

I have simplified the table in Mark Walker and David McKay. *Unravelling Genes: A Layperson's Guide to Genetic Engineering.* Sydney, Australia: Allen & Unwin, 2000, 16, for our purposes. The table appears in some form in every book cited in this chapter.

Some scientists do not like the idea of using verbs like let go, unzip, or tell because it smacks of intelligent design or the impression that these mindless entities know how to do something. On the other hand, some scientists, less resistant to the idea of intelligent design, are not bothered.[48] What we do know is that both types of scientists use the terms because there is no other way to explain what happens. No one wants to say categorically whether it's pure rote for its own sake, or there is intelligent design at work. What we can say categorically is that it is impossible to explain such terms in any other fashion than one that sounds like intelligent design. To ask someone to describe the DNA process without using words and phrases like let go, tell, and unzip is like asking someone to describe an accordion without using their hands.

As mentioned earlier, Watson and Crick noted in their paper that "It has not escaped our notice that the specific pairing we have postulated immediately suggests a possible copying mechanism for the genetic material."[49] This, coupled with McClintock's understanding of the so-called jumping gene and Paul Zamecnik's biochemical understanding of protein synthesis set in motion our complete understanding of the code of all life. The 64 codons (three are called stops, as in Table 2.2 above,) create the genetic code.

But how is it possible that all living things do not have the same code? While it may seem likely, the chances are infinitesimally small that two cells would use the same genetic code by chance. For example, suppose you are given 26 letters to represent our alphabet, only you don't know it.[50] What are the chances that you will use the symbol A for that sound that A makes? One in 26. But to correctly assign every letter to its appropriate sound, the chances are 1 in 403,291, 146,110,000,000,000,000,000 (or 1 in 4×10^{26}). The odds against two cells using the same genetic code are even greater! We'll see in Chapter 3 there is very little genetic difference between, say, a human and a chimpanzee. But that small difference is light-years wide.

Although the recombination of genes provides life's variety, it also provides for the chance of mutation (referred to earlier as exceptions). Mutations (scientists like to call them genetic variability or heritable changes) are accidents in the bases, though they are not always wrong or incorrect. In fact, there are many mutations in genes that are perfectly fine and without which we would not have managed our environment so well. Most of the time, deleterious mutations kill the organism in which they occur, effectively stopping any possible copying of that deadly gene again. In other cases, if unstable trinucleotides repeat too many times, we get something like Huntington's disease, ataxias, myotinic dystrophy, Kennedy syndrome, Tay-Sachs disease, cystic fibrosis, the dreaded Lesch-Nyhan syndrome, and more.[51] In the cases where they mutate, copy, and survive, they create a condition where the DNA will no longer be an exact replica of the original DNA.[52]

So how does the copying occur? Recall for a moment the rules by which the base sequences of DNA are copied or translated into the amino acid sequences of proteins.[53] A protein is a chain of amino acids that are themselves small organic molecules consisting, by and large, of carbon, hydrogen, oxygen, and nitrogen.[54] These amino acids are often represented as beads (and will remind those old enough to remember what were called "pop beads" that young girls played with in the 1950s). These form three-dimensional shapes. This shape proves crucial because it is this shape that allows it to match up with other similarly shaped amino acids, much like putting together a jigsaw puzzle but with instructions on every piece.

As we said, amino acids are formed in triplets of the nucleotide bases. For example, the ATG triplet is the amino acid methionine, but read backwards as GTA is the amino acid valine.[55] In order for the code to get copied, it must happen at a special location in the cell called the ribosomes (see the cell structure in Figure 2.1).[56] There it is transcribed into messenger ribonucleic acid (mRNA), known as codons (Table 2.2). Codons are the three base codes on mRNA molecules defining the amino acid produced.[57] In order to read this code, the cell goes through a process called translation whereby the cell decodes the message sent by the mRNA. For example, the code may call for brown hair and so the messenger RNA calls out, "Activate gene 3456, brown hair," and so on.[58] To be sure this code (in this case, brown hair) is carried out in a step-by-step fashion, another group of RNAs (called transfer RNAs, usually represented as tRNA) make certain everything is linked up in the manner called for by the mRNA. Crick called this process RNA transcriptions. As we said above, DNA has four bases, A, T, G, and C. It turns out that ribonucleic acid (RNA) does, too, but substitutes U (uracil) and ribose for the T (thymine). Scientists have come up with numerous analogies, calling it everything from translating a sentence from one language to another, or Morse code. The latter maybe the most apposite, as Morse code has only dots and dashes to work with. Both make up words and then sentences, much like the four bases, A, T, G, C. So, a certain sequence is read and transcribed in the cell. This DNA synthesis does not allow DNA to recreate itself from scratch. Rather, a small strand of RNA must be catalyzed by an enzyme called primase. This small piece of RNA (or primer) begins the process until it is removed and replaced by DNA.[59]

To get really technical about the process, mRNA is only one type of RNA. Another is rRNA, or ribosomal RNA, and a third is tRNA. The point is, mRNA moves about the cell to the ribosome where it meets up with tRNA and begins the process where each attaches to an amino acid.[60] Rather than coding for proteins, DNA gets a message in code for the protein that is later uncoded in the process referred to earlier as transcription and translation.

DNA is transcribed into mRNA and then translated into a protein.[61] The transcription requires enzymes devoted to this task alone (e.g., RNA polymerase). The translation of mRNA into a protein is quite complex and requires a molecule mentioned earlier, a ribosome (rRNA). The triplets are formed in a sequence (see Table 2.2), beginning with an initial codon and ending with a stop codon that ends the gene production. The entire process of gene regulation and expression ultimately winds up as a synthesis of a protein.

During this transcription process, only one strand is made. While most of the time this process is from DNA to RNA, there are occasions where it passes from DNA to DNA and RNA to RNA (in certain viruses, for example).[62] This explains why the French botanist Jean Lamarck was wrong and Darwin was right. You cannot inherit what are generally referred to as acquired characteristics, such as your father's iron-pumping physique or your mother's six-pack abdomen. These have been shaped after the fact. Of course, you do inherit the propensity to have these characteristics if you undertake the work to develop them.

As mentioned above, DNA is double-stranded (hence its name, double helix), while RNA is single-stranded. RNA is the intermediary organic compound in the synthesis of protein directed by DNA.[63] This creates what is referred to as the genetic code, or the sequence of nitrogenous bases on a DNA molecule. This code ensures the right placement of the amino acids in each cell's protein. Sequences have a 5' end and a 3' end (read as a 5-prime and a 3-prime). This is helpful to scientists because it allows them to know the direction of the sequence as well as to know how to splice it later in stem cells, as we shall see in Chapter 3.[64] In many ways, the 5' and 3' ends tell scientists which end is up, so to speak.

It took quite a while for scientists to figure out the 64 combinations that make up the 20 amino acids. If only two bases are grouped together, only 16 combinations are possible, not enough for the 20 amino acids. But if they appear in groups of threes—the triplets mentioned earlier—we get all we need, plus the three stops mentioned earlier. The stop codons are important (see Table 2.2) because they signal when the chain or copying should end. Think of these as genetic traffic signals.[65]

While the process may seem confusing to laypersons, it is mainly so because it has to be separated into many steps in order to understand it. Since there is only a certain way these can match up (A with T, G with C), it is not as confusing as it may seem. Furthermore, there are what might be called myths or urban legends about the process, and even the scientific models themselves tend to paint a picture that isn't quite right. For example, genes are not located on chromosomes.[66] Chromosomes are made up of very long molecules of

DNA. Genes do not replicate themselves, though it would seem so. Proteins replicate the genes using the gene as a blueprint or a code. While DNA is a blueprint, it is much more than even that because it not only fashions the building, so to speak, but all the tools (and trucks and laborers) to create the building! It can do all this because DNA is not so very tidy. Remember, the same DNA sequence can be read forward for one protein, but another when read in reverse.[67]

With the Watson-Crick discovery, DNA (actually D.N.A. in the paper) became a household word overnight. Because of the discovery (and subsequent splicing of genes elaborated on in Chapter 3), we now see possible cures for cancer, the elimination of diabetes, and the guilt (or innocence) of the rich and famous, as in the case of O. J. Simpson.[68] At least that is our hope. But how can it be possible that something so infinitesimally small can be responsible for so much? Bear in mind that every cell in the human body has a strand of DNA. It is estimated that each human has about 6×10^{12} meters of DNA, or enough to reach the moon and back 8,000 times if laid end-to-end.[69] Each cell has 46 chromosome pairs with about 150 million base (A–T, G–C) pairs. It is hard to imagine that four bases (A, T, G, C), combined three at a time, yield enough to complete the 20 amino acids found in all proteins. Yet the $4 \times 4 \times 4 = 64$ provides the right amount.[70] This endless repetition allows for the continuation of the code of life. What is even more amazing is that you and I do not differ all that much, as each of us has about 25,000 different genes. What causes us not to be identical (unless we are twins, and even then there are some differences) are variations called polymorphisms that occur once in every 1,000 bases. Each of us has about 3 billion bases of DNA, meaning there are 3 million differences between you and me.[71] This is why DNA fingerprinting in police work has risen to such importance (see Chapter 8 for further discussion).

Of course, all this work took more than Watson and Crick to accomplish, even though they are undoubtedly what we might call the superstars, the million-dollar players (literally, after winning the Nobel Prize). But those who went before them, such as Delbruck and Miescher, and those who worked along side of them, like Franklin and Wilkins, deserve much credit as well. The discovery of DNA has as much an historical flavor about it as it does an international component. Hard as it may seem to those of us living in the twenty-first century, these men and women, as mentioned before, did all this very complex work without the technological amenities we now enjoy. Much of the work came about through collaboration, hard, long hours of study and observation, and endless repetition, failure, resumption, failure, continuation, failure, review, and finally success. This element of arduous scientific experimentation cannot be overlooked for it remains, even today in our highly technological society, a key ingredient to important scientific discoveries. Indeed, over-reliance on computerized research alone can lead to complete failure, or even death.[72]

Consider Bruce Fraser, almost overlooked in this genetic history. Fraser had written an unpublished but seminal paper on DNA just before Watson and Crick discovered the double helix and its properties. In fact, they cited it as "unpublished experimental results and ideas" in the last paragraph before their notes. It was not in press and indeed had not been submitted to a journal at all because Mr. Fraser, a postgraduate assistant, showed it to the graduate director of his program who told him publication was premature. Fifty-one years later, when he was 80 years old, Mr. Fraser's paper was published in the *Journal of Structural Biology* in 2004.[73]

Equally overlooked were two women researchers, Barbara McClintock and Harriet Creighton, who published a series of studies in 1931 proving that genetic information crossed over during cell division, a process technically known as meiosis.[74] McClintock tried to find out why these so-called jumping genes existed, these genes that seemed to move from their original position to another, new one. She discovered this by using corn.[75] The process came to be known as recombination and led to the now well-known process of recombinant DNA that opened the door for stem cell research. McClintock was belatedly awarded the Nobel Prize in Physiology in 1983.[76] These vastly unsung heroes and heroines need to be recalled in any discussion of DNA to be sure they will not be overlooked again. They represent scores of others without whose work we would still be fumbling about, trying to explain this miraculously astonishing process.

Bear in mind that this simple system of four units (really subunits) can produce a human brain of millions and millions of interconnections, a human heart that pumps ceaselessly 100,000 times a day (perhaps for as long as a century), a human immune system that can attack foreign bodies while not killing itself, and a human reproductive system that can replicate yet another one of these miraculous organisms all over again; DNA relies only upon length.[77] In Psalm 139:14, David wrote, "For I am fearfully and wonderfully made." and he didn't know a fraction of what we know today. Regardless of one's religious disposition, it's hard to look at this process and not be similarly amazed.

The discovery of DNA gave rise to the possibility of genetic engineering, or the process of directly manipulating the genome of an organism for a desired end.[78] Because this process requires combining two or more genes from two or more sources, it is also called recombinant DNA (rDNA) technology. It is in this process that all controversy resides.

For example, not many of us would be bothered by a rubber mouse.[79] We know it to be fake. It never tries to pass itself off as a real mouse, though we might, in a mischievous mood, leave it about to scare someone. What would bother some of us would be a mouse created in some laboratory that really did bounce! We're troubled by creations that seem to be made for no other purpose than to prove that they can be created (recall our myths that began this chapter). These are not real and they seem to serve no real purpose. Not many of us would oppose genetic engineering to save lives but more than half of us would object to genetic engineering to save some lives at the cost of taking the lives of others or create endless chimeras just for the sake of creating them.

During Hitler's reign of terror, many scientists performed all sorts of terrible and deadly experiments on Jews because they saw the Jews as superfluous, as only half-human or at least not fully human.[80] Hardly anyone can find an ethical reason for these experiments, which were really nothing short of torture. Before we feel too smug, however, we need only to consider our own government-run program at Tuskegee Institute in 1932 in which a number of African American men with syphilis were purposely left untreated to determine the course of that disease, even to their deaths. That program was no accident; it was implemented with that outcome planned in advance.[81]

This attitude became even more pronounced once DNA and its attending industry were launched. While many benefits accrued to many scientists, the recombinant DNA techniques, as they would be called, opened the door to many possible medical benefits as well as numerous biohazards.[82] It's not too hard to see why. Almost at once, many in the scientific community were struck with the possible applications in human society: agriculture, wildlife management, animal husbandry, food preparation, the world's food pantry, manufacturing, and an endless list of opportunities in medicine with the potential to cure everything from high blood pressure to allergies to nearly every cancer known to man.[83] Man had indeed found the key to playing God. But did he, like Adam, really want to know too much and hold a gift too terrible to control? Like Pandora or Icarus, did he now have that which he could not properly direct? Many thought so, and this sentiment led to the Asilomar Conference of 1975, about which more is presented in the next chapter. For now, it is enough to say that not a few scientists were scared witless about the possibilities.

The discovery of DNA was not heralded in every circle. Until the discovery of DNA, it was easy to think of dinosaurs as extinct when other, more familiar forms of mammals began to appear. But recent DNA studies show that many mammals coexisted with dinosaurs. That is, their DNA matches that of organisms that could not possibly have existed in that time frame, or so we thought. Paleontologists have referred to this theory of evolution as maddening.[84]

Indeed, the whole field of biogenetic engineering has changed, for good or ill, our understanding of what the words natural and unnatural mean. As biologist Michael Fox points out, "Genetic engineering makes it possible to breach the genetic boundaries that normally separate the genetic material of totally unrelated species."[85] As techniques for genetic manipulation have become more and more sophisticated, more numerous and more unnerving (at least to some) definitions of what is normal have evolved.

By the mid-1990s (1996, to be exact) scientists in Great Britain had abnormally created (cloned) the first mammal, Dolly the sheep (this will be further discussed in Chapter 5). A year later, scientists in Hawaii cloned mice and then made clones of these clones. Following that, scientists reported the first stem cells (to be examined in Chapter 3) cultured from human embryos.

In 2001, a great deal of genetic engineering occurred. The U.S. Congress rushed to pass a bill banning all human cloning. Later that year, President George W. Bush allowed federal funding for research on human embryonic stem cells, but only on existing cells, not new lines. Still later, Advanced Cell Technology, a company that had been trying to clone human embryos, announced that while it could get human embryonic clones, they stopped dividing before stem cells emerged. The year ended with the birth of the first cloned household pet, CC the kitten.

In the spring of 2002, work on mice with immune disease and cloned transplant cells suggested that so-called therapeutic cloning just might work. While some would call all this wonderful, something strange also occurred in that particular year. One company linked with a UFO religion claimed to have cloned a baby girl. No proof was offered and the claim was later proved to be bogus. In 2003, cloning attempts on monkeys failed miserably and provided some evidence that the technique used on Dolly might not work on humans after all. In late 2003, the United Nations delayed a vote on a ban on all human cloning at least for two years. In February of 2004, Panos Zavos, an infertility physician, reported implanting a cloned human embryo into a human womb but with no results. That same month a Korean group claimed to have 30 cloned human embryos, with one yielding embryonic stem cells. In August 2005, Korean researchers claimed the first cloned dog, "Snuppy."[86]

All of these discoveries created important but as yet not completely resolved ethical questions. All too often the scientific technology has developed long before the ethical issues have been addressed or even discussed. But it isn't as if we haven't been down a very similar road before. The question remains as to whether we have learned anything from that visitation.

A BRIEF HISTORY OF EUGENICS

From about 1910 through the 1940s, Americans were concerned—terrified might be the more appropriate word—by what might be called the menace of the moron.[87] Before the discovery of DNA, and only just after we had come to understand the power of heredity, Americans were quite aware of their genetic pool and they wanted to keep it clean. The concern, or the terror, went under the unassuming name of eugenics, the science of race improvement through the control of hereditary factors. That usually meant encouraging so-called good breeders and hindering the rest by any means necessary. As one researcher has put it, "The revolution in genetics, although full of promise for understanding our own constitution and for the power to change human lives for the better, has nevertheless proven profoundly unsettling."[88]

If Americans had been aware of where the idea of a pure or better race that eugenics aspired to had come from, perhaps they would have been less enthusiastic. Friedrich Nietzsche, the German philosopher, had already sketched the idea of the Overman and the concept of selective breeding. As early as 1885, he outlined the importance of separating the so-called master race from what he called contamination from the weak, "a race with its own vital sphere ... a positive race which can enjoy every luxury."[89] As with virtually any call for race betterment through anything other than personal improvement (either by morals, education, or both), this sentiment later inevitably led to the terror of the Nazi regime and the Jewish Holocaust.

American textbooks touting eugenics and the unseen terror of a weak genetic pool alarmed readers thusly:

The proportion of those who are feeble-minded in such various directions as to constitute the feeble-minded class is estimated at 3 percent of our population, and were we to include drunkards, paupers, grave sex-offenders, the criminalistic, the insane, and those with innate physical weaknesses that render them for the most part incompetent, it seems a safe estimate that 8 percent of our population are far from having the capacities of effective men and women able, not merely to support themselves, but really to push forward the world's work.[90]

It sounded like an epidemic, though it never (in America, anyway) amounted to more than 3 percent of the population. When the army recruits of the First World War were tested, nearly 2 million were found to be barely literate. Scientists claimed these tests measured native intelligence, not education or training. But scientists at the very core of the profession—men like Havelock Ellis, the renowned scientist who set for a number of generations our understanding (some would argue, our misunderstanding) of sexual relations—stated in no uncertain terms what must be done. Calls to "control the stream at its source" and "prevent the contamination of that stream by filth" and not allow it to "muddy" up our race and sweep our hard work away, brought eugenics into the mainstream.[91] That filth, by the way, was our fellow Americans! Indeed, rereading some of these reformers (including preachers from their pulpits) gives us pause as we learn, for instance, that the children of prostitutes are so undeserving that death is "nature's great blessing" to them.[92]

To add fuel to this fire, Darwin's theory of evolution made eugenics appear even more respectable.[93] Eugenics had been pushed to the forefront of the discussion and, whether eugenicists were reform or mainline (the classification is Kevles'), the outcome proved identical: clean up the genetic pool.[94]

Perhaps the scare rose from the vast number of immigrants who flooded into this country, those "tired, poor, huddled masses" that our Statue of Liberty welcomes to freedom. Others were even more bare-knuckled than even Ellis. H. J. Muller claimed that "imbeciles should be sterilized ... [that] is of course unquestionable." Had the tool of genetic selection and screening (see Chapter 7) been available, it would have no doubt been required.[95] Even Raymond B. Fosdick, designer and president of the well-known Rockefeller Foundation (1936–1948) urgently called for "fearless engineering in the social field" that would push the bounds of human understanding.[96] Francis Galton (1822–1911), a most prominent scientist who later helped develop intelligence tests still used today and known for his groundbreaking work in statistics, argued that while almost nothing was known of the laws of heredity, we could know which parents we wanted to breed and which not.[97]

Galton's work proved especially pivotal. Aping Mendel, Galton used sweet-pea seeds "as ... anthropological evidence ... as [a] means of throwing light on heredity in man."[98] Coupling this data with data from the Anthropometric Laboratory and the *Record of Family Faculties,* Galton was able to show or, at the very least, suggest the hereditary outcome of certain families. By showing things like height, weight, hair color, and so on, he could infer intelligence with little trouble. This evidence, coupled with the evidence that I. Q. (evidence that army recruits later provided) proved ultimately compelling: if we wanted to keep the race clean (i.e., pure), there would have to be selective breeding.[99]

Scientists not only had help from other scientists and the clergy but also giants like America's own birth control doyenne, Margaret Sanger, and likewise London's own Mary Stopes. Both women strongly urged eugenic methods that were legally enforceable.[100] George Bernard Shaw's *Man and Superman* was a veritable paean to the new race of superbreeders. Jack Tanner, the play's anarchist, blatantly lists what science can already do with animals and declares, "[W]hat can be done with the wolf can be done with man," meaning, of course, if we control breeding, we can breed a race of supermen.

Mendel's work became most important and little tableaux, called Mendel's Theaters, were circulated about the country to show that if brown hair and blue eyes could be passed down, why not shiftlessness and indolence?[101] The most unfortunate families of Ada Juke and Martin Kallikak became cases in point. Ada Juke's genealogical line was marked by an astonishing number of so-called morons and imbeciles. Pictures of her and her family were everywhere, warning that "Most social problems are produced by comparatively few families; the elimination of a few defective lines would result in an enormous reduction in our public bills."[102] It did not stop there. Families were rated by their ability to produce the right kind of offspring. If America wanted to keep its pure line, sermons must be preached (and they were) and young men warned that they must not ever stray but stick to their own kind. Indeed, if America was to remain strong, if she was to minimize human suffering and maximize the public good, she must make certain her young men did not follow what was euphemistically called blind instinct. If they did, our country would be as corrupted as Icarus and would fall as he did.[103] This led to fairs and parades of the so-called right kinds of families. Note carefully the altruistic arguments ("for the good of all," "to prevent disease," and "reduce public bills") that make these reformers enthusiastic. We'll see them again in Chapter 7.

These scientists were not without challengers. The great novelist Sinclair Lewis mocked the movement in his *Arrowsmith,* in which the main character's family, a so-called eugenic family at the Iowa state fair, is presented as the "Holton Gang"— the parents unmarried and the children far from perfect. G. K. Chesterton, the great English essayist, denounced the movement as part of what he called scientific officialism and even Prussianism, while the Catholic Church denounced sterilization.[104] It did not matter. These few voices cried in the deserted darkness. Middle America, especially as represented by Midwesterners, found these arguments against a weak gene pool most convincing. Add to this evidence, however shaky it was, the Fitter Families Contests at county fairs in Iowa, Kansas, and elsewhere, and a veritable pure race movement had begun, right here in River City. There was nothing subtle about these fairs, either. Fairs broadcast on posters that "pure + pure = normal children." But "abnormal + abnormal" or even "pure + abnormal" meant either an impure outcome or a tainted one.[105] Was it this desire to breed pure lines that slowed America's response to Hitler's madness in World War II? We cannot say with certainty, but surely eugenics and our embrace of it in this country caused some to wonder what all the fuss over Hitler was about.

Perhaps eugenics would prove to be just another dream of the intellectuals and nothing more. More often than not, far-from-the-mainstream ideas issued from academia (and this continues today) have, thankfully, little or no effect on the general public. Eugenics could be the same, could it not? In a word, no. Dozens of sterilization laws were passed by states, such as the one adopted by Virginia in 1924. Its effects were hardly innocent. By 1958 more than 61,000 Americans had been sterilized, 20,000 in California alone.[106] In the early

1920s, when the 15-year-old pauper Betty Stark of Lynchburg, Virginia, turned up pregnant, she ended up in a facility for the mentally retarded at her father's request following the delivery and subsequent adoption of her child.[107] Her I.Q. test revealed an intelligence quotient of 72, bordering on low retardation. Her physicians scheduled her for an appendectomy that Betty learned only many years later, and following much grief, was in fact her sterilization. None other than Supreme Court Justice Oliver Wendell Holmes wrote the decision upholding the constitutionality of the sterilization in *Buck vs. Bell,* believing that Emma Buck and her daughter Carrie should no longer have children. Another daughter, Vivian, was also sterilized, though the only so-called proof of her unsuitability was her lineage. Her intelligence was never tested.[108] Holmes's "Three generations of imbeciles are enough," which he wrote in the final decision, has been the rallying cry for continued genetic selection and remains perhaps all that is remembered of that doleful decision. Julian Huxley added more credibility, arguing "The human species is in desperate need of genetic improvement."[109]

Some may view this closing discussion on eugenics as nothing more than a sidebar to a book on genetic engineering. But it serves as much more than that. Although the founders of eugenics can be readily dismissed as overzealous or worse, many genetic engineering proponents today still refer to positive and negative eugenics.[110] That is, there are certain kinds of genetic manipulation that have negative side effects (for instance, the sterilization laws mentioned above) and certain kinds that are positive (markers for disease). But as we shall see, it isn't as easy as that. Any time anyone, however smart he or she may be, begins talking about the betterment of the race at the expense of others, bells must go off and alarms must sound. As long as one is on the benefiting side of the equation, one may not see the danger. But slide over into the downside and it suddenly all becomes very clear. Think only of the masses and forget the individual, and all is lost.[111]

The idea behind eugenics—and some argue that it surfaces from time to time even today—proves that even uncertain scientific knowledge in the hands of experts can be wrongly used. While there are some who contend that eugenics should not be dismissed, especially in the context of new scientific developments that make its goals all the more achievable, others are quite certain that the eugenics episode in our history should serve only as a parable warning of the dangers of overreach.

At the very least, it should give all of us pause as we enter into a new age that provides even more technical abilities to bring to pass a race that is stronger and better. As we shall see in Chapter 7, genetic engineering allows parents to choose not only the gender of their offspring but also how that gender will be expressed. In the wrong hands, genetic engineering allows parents to select males over females while eliminating what we define as undesirable: the weak, the slow-witted, the less able, and the differently abled, including persons with Down's Syndrome. To state it another way, are we ready to distinguish, as one writer has put it, between the enhanced and the unenhanced?[112] Is this what we want? Is this what we should want? Once we begin this process (and many say it has already begun), can we ever go back? If parents are allowed to choose gender, will they also be allowed to choose even more (as for example, heterosexuality over the slimmest possibility of homosexuality, assuming that that gene can indeed be identified)?[113]

Genetic engineering has opened a Pandora's box. It has given scientists the power of so many Daedaluses for any number of lesser Icaruses. Whether we select for gender or select for the best, or select to erase part of the population through biological genetics, the genie is out of the bottle. Now that we have opened the box, have we reserved enough hope for future generations? Will we prove the fire in the lap of God prophetic? It is to these controversies that we now turn.

3

Splicing, Dicing, and Cloning to Asilomar and Beyond

If Mary Shelley's *Frankenstein* surprises us with its foresight in predicting mankind's insatiable desire to be too much like God (creating life asexually, so to say),[1] Aldous Huxley's *Brave New World* shocks us with uncanny prescience describing, long before it could be known, a production facility that makes humans. We pick up the story at the London Hatchery where its director is lecturing delighted students about the fertilization process. It bears a lengthy recitation:

"The week's supply of ova. Kept," he explained, "at blood heat; whereas male gametes," and here he opened another door, "they have to be kept at thirty-five instead of thirty-seven. Full blood heat sterilizes." Rams wrapped in theremogene beget no lambs.

Still leaning against the incubators he gave them, while pencils scurried illegibly across the pages, a brief description of the modern fertilization process. . . . [He] continued with some account of the technique for preserving the excised ovary alive and actively developing. . . . [T]his receptacle was immersed in a warm bouillon containing free-swimming spermatozoa—at a minimum concentration of one hundred thousand per cubic centimeter. . . . [T]he fertilized ova went back to the incubators . . . where the Alphas and Betas remained until definitely bottled; while the Gammas, Deltas, and Epsilons were brought out again, only after thirty-six hours, to undergo Bokanovsky's Process.

"Bokanovsky's Process," repeated the Director, and the students underlined the words in their little notebooks.

One egg, one embryo, one adult—normality. But a bokanovskified egg will bud, and every bud will proliferate, will divide. From eight to ninety-six buds, every bud will

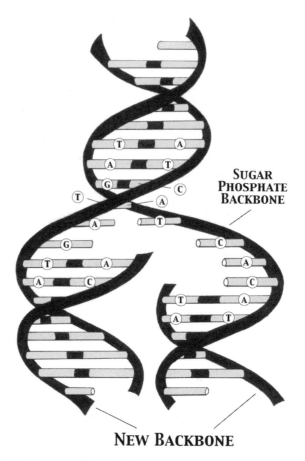

Figure 3.1   DNA unzipping

grow into a perfectly formed embryo, and every embryo into a full-sized adult. Making
ninety-six human beings grow where only one grew before. Progress.[2]

Of course, looking back now after so much science and modernization,
Huxley seems dated, but only slightly. Bear in mind that the book first
appeared in 1939. Astonishingly, before the double helix discovery and stem
cell research, Huxley had pointed out the ethical and moral quandary. Huxley,
a brilliant mind, more or less described the process of gene splicing, stem
cells, and genetic engineering long before they ever became even household
words among scientists. *Brave New World* is the new Eden, brought to us
courtesy of the new science, the new messiahs who are able to bring us everything we want and more. In it, Huxley chronicles the fictive life we are freed
from—the chains of patriotism, religion, morality, and all the rest that puta-

tively enslaves us. But we give up one thing for another without knowing if we have made the better choice. The question today is, is it all still fiction?

Genetic engineering has brought us a new world but at what price? Have we become the "tools of our tools," as Thoreau warned against? Have we become, in this new world, soulless creatures without personalities? According to C. S. Lewis, "If any one age really attains, by eugenics and scientific education, the power to make its descendants what it pleases, all men who live after it are patients of that power. They are weaker, not stronger ... [for] if the dreams of the scientific planners are realized ... the rule of a few hundreds of men over billions upon billions of men [has been vouchsafed]."[3]

Is this too harsh an assessment? Nothing in science actively seeks to impose a soulless, zombie-like existence on man. And yet there is an uncanny resemblance in our technology today, and the technology of the Brave New World, to deliver what some could consider the very world of Huxley or the very "men without chests" of Lewis. This is less a value judgment than a recitation of the history itself. Can we control what we have learned, to our best good?

AFTER THE HELIX, THE FLOOD? SPLICING AND DICING GENES

Mendel's work explained how traits were passed down. As we saw in Chapter 2, this was refined and explored until it was fully explicated by the Watson-Crick double helix model. That discovery led to genetic engineering, a general term that can also refer to DNA cloning or recombinant DNA.[4] Genetic engineering is "the deliberate manipulation or alteration of the genetic material of the cell, in order to directly produce a change in some feature of the organism."[5] Cloning, from Greek *klon* that means bud or twig, refers to a deliberate genetic manipulation with an end result being genetically identical material.

It all sounds innocent enough until one realizes that changing "some feature of the organism" means changing some feature in you or me. That can be a good or bad thing, depending on which feature is changed, whether the change is permanent, and whether the change is with or without one's permission. The eugenics discussion at the end of Chapter 2 provides ample evidence that changing a person's features without his or her permission, even in America, is not impossible.

So just how is this change made? Recall from Chapter 2 the unzipping of the double helix, whereby DNA, with the help of transcription and translation processes, creates genes. Once these processes were understood, scientists began to tinker with them.[6]

Technically, this tinkering, called stem cell transplantation, is the process of covalent joining (known as ligation) of any piece of DNA (referred to as an insert) that is contained in an unresolved mixture to a replication-competent and selectable DNA element (called a molecular vector). These are then transferred into host cells.[7] In laymen's terms, this means taking a piece of DNA that can be separated, cutting it out, and placing it into another host. The first such experiment (performed by Peter Lobban and Dale Kaiser at Stanford University) used ribosomal RNA genes (described in Chapter 2) from a frog and placed them into a bacterium, such as *Escherichia coli* (or *E. coli* for short). *E. coli* is a widely used and readily available host for scientists working in this field.[8] However, Paul Berg is generally credited with creating the first recombinant DNA molecule, while Herbert Boyer and Stanley Cohen in 1972 are credited with refining the actual process to the one used today.[9]

CUTTING REQUIRES SCISSORS; PUTTING IT BACK REQUIRES TAPE

But none of this was possible until about 20 years after the Crick-Watson discovery. In 1970, Hamilton Smith and Daniel Nathans, both of Johns Hopkins University, discovered a class of what are now known as restriction enzymes. These could be used as scissors to sever DNA at specific spots, making it easier to cut and splice, so to say, where desired.[10] This process really revolutionized genetic engineering, allowing for all sorts of DNA sequence cuttings to be made and what was left to be inserted in the same or even different organisms. This cutting and splicing has been compared to many things to explain what is taking place; arguably, the best and most vivid analogy might be the editing of film. While the cutting tools are the restriction enzymes, the splicing tools are another class of enzymes called ligases.[11] Ligases connect the ends of DNA segments and seal or glue them together.

In effect, then, there is one tool, comparable to scissors, which cuts the DNA, and another tool, comparable to tape, which seals it back together. This may help explain why the DNA from a bacterium can be linked to animal DNA. Of course, herein lies part of the controversy. Scientists complain that the public misunderstands this process (undoubtedly true) and so should not fear mouse-headed fish coming out of laboratories. On the other hand, if scientists can connect bacteria to animals, why not worry about mice-headed fish? In fact, a mouse with human DNA does exist, not to mention mice that grow human-like ears.[12] The answer resides at the cellular as well as the practical levels.

DNA SPLICING

This cutting and splicing process, coupled with various successful experiments, immediately showed scientists that genes could be produced for study

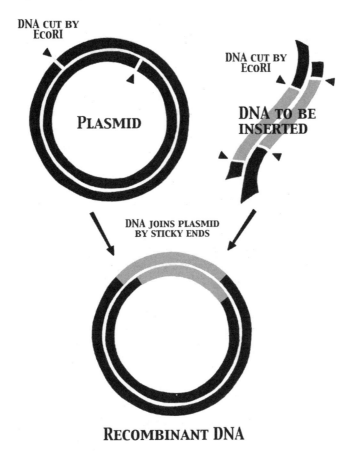

Figure 3.2   Gene splicing

and for industrial uses. Later, the process was transferred to plants (to be discussed in Chapter 4) and animals (Chapter 5), giving rise to the creation of what are called transgenic organisms. What scientists discovered is that this process allows the host to replicate in some or all cells and the resulting organism then passes this new code on to succeeding generations. While the process is fairly easy, what has not been as easy is the targeting of certain genes (say, defective ones) and replacing them with a nondefective ones. This technique led to the establishment of collections of DNA fragments into what are called genomic libraries. These libraries provide for materials to be used for sequencing, for other DNA study, or for other kinds of cloning projects and applications. The collection of these libraries led to the Human Genome Project (described in Chapter 6), which involved the mapping of all 46 chromosomes, what they do, and where they reside.

STEPS IN SPLICING

While it is not the purpose of this book to provide minute details of the splicing process, some general idea will be helpful in understanding the relevant controversies to be discussed later. The steps involved are the insert preparation, ligation, transformation, selection of transformants, and the screening of the clones.[13] As one of the great developers of recombination tools, Stanley Cohen, put it, "There are four essential elements: a method of breaking and joining DNA molecules from different sources; a suitable gene carrier that can replicate both itself and a foreign DNA segment linked to it; a means of introducing the composite DNA . . . and a method of selecting from a large population of cells, a clone recipient."[14] The insert is prepared by collecting large amounts of the DNA fragment by digesting genomic DNA with a restriction enzyme. Most restriction enzymes recognize short sequences that are palindromic. These are split in a staggered way, yielding a few hundred to a few thousand fragments. Because scientists know how DNA connects (described in Chapter 2) and are familiar with the 20 amino acids (along with the stop codons), this process is not as mind-boggling as it seems.

The most commonly used vectors are plasmids, or extrachromosomal elements, that are able to replicate themselves within their hosts apart from the host chromosome. Large numbers of these vectors come from bacterial plasmids, especially in *E. coli*; hence, its common use in laboratories. Another commonly used vector is the phage (shorthand for bacteriophage, described in more detail below) referred to as lambda, or the plasmid found in yeast. The ligation step is accomplished by another phage (called T4) prepared from cells infected with *E. coli*. The transformation step introduces this into the host. Since this does not occur naturally (except in viruses), it must occur via chemical or physical treatments that make the host able to accept them. In fact, the process in viruses mimics what happens in the transformation process. A virus (called a bacteriophage or phage) attacks the bacterial cell membrane and inserts its DNA; the DNA is then replicated (both its own and the cell's) and later leaves. When it attacks another bacterium cell, both the virus and the cell's DNA are replicated into the new bacterial cell, and so on, releasing the virus vectors throughout the organism.[15]

Transformation has a low yield, so the selection of transformants is critical. Bacteria like *E. coli* and yeast provide the best and often easiest conditions for this to take place. Of course, more complex organisms require a more highly sophisticated technique (beyond the scope of this book), but this provides a commonly accepted, general overview. From a scientific view, it just keeps getting better. At the close of the last decade, geneticists were excited over the discovery and manipulation of so-called infectious clones or "cDNA copies of entire viral genomes in bacterial plasmids in which either the DNA itself or the RNA transcribed from the cDNA in vitro is infectious."[16]

Gene splicing has been labeled as the most important physiological technique to be developed in the past few decades. The process of isolating a piece of DNA—a sequence—including one or more intact genes and inserting them in a functional way into another organism, sometimes even in an unrelated organism, has enabled genetic engineering to hold such future promise.[17] Gene splicing is often seen by the lay public as a separate science but it is really only a technique used by scientists to achieve certain results. The results can be remarkably beneficial; they can also be extremely controversial.

RECOMBINANT DNA AND POLITICS

This controversy is unfortunate because a few celebrated discoveries either elicited a great presumption and expectation about all the rest or cast a dark shadow of doubt about the future of many others. In the 2004 presidential campaign, the Kerry-Edwards campaign made much about stem cell research, a legitimate campaign debate topic. But the hope with which it was discussed far outstripped what we know today or even think we can know 20 years from now. Debating in this manner causes sides to be pitted against each other for political reasons at the expense of scientific progress, however the latter is defined. Unfortunately, this kind of debate has always hounded genetic engineering, as even the terms we use to discuss it often lead to confusion.

For example, terms like clone and even test-tube baby are alike in their ability to mislead the public while capturing the headlines. We are stuck, it seems, with having to couch the debate in brave new world terms or in language so foreboding as to conjure images of Frankenstein. The word clone, for example, as Burke Zimmerman points out, is really only "a collection of identical copies from a single entity."[18] As was shown in Chapter 1, however, popular media—journalism and Hollywood—turn it into a horror story as in *The Island of Dr. Moreau*, where monstrous creatures, half-animal, half-human, roam freely. Another example is the movie *The Boys from Brazil*, where science produces dozens of carbon-copied Adolph Hitlers. If fear doesn't get you, then science fiction will. Readers over the age of 40 will recall a television show that began with a gruesome accident that left a man barely alive. Then a commanding, confident voiceover said, without hesitation, "Gentlemen, we can rebuild him. We have the technology." Thus was born *The Six Million Dollar Man*.[19] He was soon followed by a woman of similar cost and it was all great fun. Today, it still remains fun to many people but also a fearful reality to others.

SEND IN THE CLONES

Cloning presents us with a "chip off the old block," as one scientist has put it.[20] This is all well and good when the old block is the old man, the father. But when the father is in a test tube or a petri dish, it begins to lose its neutrality. The evolution of the word cloning has pushed its meaning from a plant asexually generated to all asexually propagated organisms. This notion of mass-producing virtually any organism is awash in debate. Although it portends great hope for the future of the control or even elimination of disease (not to mention the salvation of species at or near extinction),[21] it also disturbs many with the fear of "at what cost?" Moreover, recombinant DNA produced organisms that would not otherwise be found in nature. Were they all benign? That was the question on the minds of some scientists.

LET THE SUNSHINE IN: THE AGE OF ASILOMAR

The ethical and moral controversies began to be better understood by early 1970. Call it the Age of Aquarius. Call it Vietnam. Call it the Age of Protest. Perhaps, as some have indicated, it grew out of the hippie phenomenon of the late 1960s.[22] Whatever the reason, the debate about recombinant DNA emerged among the scientists doing the work in the early 1970s.[23] By 1973, scientists were clearly beginning to have second thoughts. Today, hardly a newspaper is printed without some story celebrating (or warning about) some new discovery. Most discoveries hold remarkable potential for the cure of intractable diseases. Others may offer no immediate advantage but may portend great advantages in a few years, or even decades. Much of this genetic manipulation overshadows what we have come to understand about cell biology.[24] But perhaps because it was so misunderstood, the very idea frightened many.

Some experiments in the early 1970s clearly involved potentially hazardous materials performed with none of today's safety protocols in place. Was it safe merely to wash away experimental residues down so many lab sinks? Scientists at the 1973 Gordon Conference on Nucleic Acids said careful note should be taken of the rapid progress being made in DNA-splicing methods and whether or not they could lead to the creation of biologically hazardous materials (they could). Indeed, some scientists called for a discussion of the "potential dangers inherent in the new technology."[25] In the early 1970s, Paul Berg of Stanford University had developed a very useful and somewhat easy technique of joining DNA molecules.[26] Berg, who played a key role in that Gordon Conference, had been working with a particular gene, SV40, and *E. coli*. SV40 did hold some dangerous, tumor-provoking potential.[27] Apart from their scientific

Scientists discuss Asilomar recommendations, Asilomar Conference, 1975. From the National Library of Medicine, History of Medicine Collection.

lure, the 1971 National Cancer Act initiated by President Richard Nixon and passed by the U.S. Congress had also spurred interest in these experiments.[28]

Another scientist, Robert Pollack, had been working with James Watson's staff at Cold Spring Harbor Laboratory and had also been studying SV40; he had seen how it could cause tumors in small animals (though none had been known to cause cancer in any human).[29] Dr. Andrew M. Lewis, working in yet another lab, also saw the danger in working with SV40. Berg's SV40 experiments alarmed them (and others) because of the possibility of producing an organism that could carry a tumor virus into humans, with an even higher risk to children.[30] Again, it is important to stress that only the possibility—no actual evidence—was present. However, it remained a serious possibility.

Pollack was quite alarmed because he knew of Berg's plan to introduce the DNA of SV40 into a well-known virus, lambda bacteriophage, and allow it to carry the SV40 hybrid into *E. coli*.[31] Pollack called Berg and told him of his concerns. Berg was at first nonplussed but he promised to think about it. He could, he told Pollack, turn the genes "on and off" at will, and so, not to worry.

Of course there was no external incentive for Berg to do anything. The research was groundbreaking, very new, and held huge industrial (i.e., financial) potential, not to mention great scientific prestige and allure. Then there was that national war on cancer (see below) that Congress had just declared, providing huge sums of money for the very thing Berg was doing. Pollack and Berg weren't the only ones working on SV40 or any of a number of similar experiments. Raising alarms, especially when unnecessary, would not affect merely Berg or Pollack but all the scientists working in the field. Indeed, when word got out that there might be a reason to hold up the research, not everyone was happy.[32]

BELL-BOTTOMS, BAD HAIR, AND DNA RESEARCH GUIDELINES

By the summer of 1974, Berg had had time to think beyond the laboratory and what such work might mean for mankind. Apparently Pollack saw the dangers before others, or perhaps he was just the first to think seriously about what he and others were doing. Following the 1973 Gordon Conference, the National Academy of Sciences appointed a committee headed by Berg. He later sent a letter to *Science* magazine, outlining the committee's recommendations about what should and should not be done with respect to the manipulation of genes. Berg called for a voluntary moratorium on experiments that: (a) might introduce bacterial toxins or antibiotic resistance into strains or sequences that did not contain them; (b) create bridges between DNA from oncogenic (tumor-producing) or animal viruses to bacterial plasmids or any other viral DNAs; or (c) weigh very carefully any plans to link animal bacterial plasmid DNA and bacterial virus DNA.[33] Since *E. coli,* a common bacterium found in the colon, urinary tract, and upper respiratory passages of humans, was the sought-after host for many of the DNA hybridizations, it seemed logical that if any escaped it could easily find its way into humans; hence, the need for the moratorium.[34]

These recommendations caught the eye of the federal government, and the Director of the National Institutes of Health (NIH), leading to the establishment of an advisory committee to oversee three kinds of DNA experimentations. It would: (1) oversee and evaluate potential biological and ecological hazards of the types of recombinant DNA molecules mentioned in the Berg recommendations; (2) develop procedures that would minimize the spread of such molecules in human populations; and (3) devise safety guidelines to be followed by all those working with potentially hazardous recombinant DNA molecules.

Points 2 and 3 became a reality with the NIH Recombinant DNA Molecule Program Advisory Committee (RAC for short) in 1975.[35] It would provide guidance on procedures that would minimize the spread of such molecules and guidelines (and advice) for those working with potentially hazardous recombinant DNA. The NIH did not involve itself to oversee the risk assessment program. The NIH did finally conduct a few experiments on weakened *E. coli* strains to transfer tumor virus genes to animal hosts.

The Berg letter contained a final recommendation for an international conference of scientists involved in research of recombinant DNA. That conference became a household word among these scientists, many of whom came to it with the bad hair, long sideburns, and bell-bottoms that were plentiful at the time. Today it is referred to generally by where it was held, in a conference center in Pacific Grove, California in February, 1975. It remains the touchstone event in the development of safety procedures with respect to genetic manipulation: the Asilomar Conference.[36]

THE ASILOMAR CONFERENCE

The Asilomar Conference brought together large numbers of international scientists to discuss recombinant DNA. A moratorium had already been called for on research involving potentially hazardous materials. A simple Internet search bears out this history. Up until 1975, a steady stream of articles were published on genetic engineering, with only a handful in the early 1960s but clearly a sudden uptick by decade's end. In the early 1970s, published research doubled and tripled with each passing year. In 1974 and 1975 it all but stopped. Clearly something important had occurred. Following Berg's letter and the subsequent NIH guidelines, published research began again.[37] It has grown almost exponentially ever since.

The Asilomar Conference showcased a number of important scientists, Berg chief among them. What constituted hazardous research to some was completely discounted by others. For example, Roy Curtiss III (University of Alabama Medical Center, Birmingham) read Berg's letter in the magazine *Science* calling for the moratorium and immediately ended his experiments using *E. coli* for a dentifrice that would retard tooth decay.[38] Curtiss did more than end his experiments; he wrote up his reasons in a 16-page, single-spaced letter and mailed it to more than a thousand scientists. Needless to say, his letter sparked interest and controversy at Asilomar. Michael Oxman argued that while a scientist may wish to take certain risks with an experiment, he cannot ask others—graduate assistants, secretaries, and the uninvolved and uninformed public—to take them as well.[39] His concerns were never really addressed.

Some scientists called for federal intervention, while others said it should be the responsibility of each scientist, not the government's. Most worried that government regulation would lead to the cessation of research in one area and an overconcentration in others. Not a few worried that grant money would be driven by so-called acceptable research while legitimate but less acceptable research would cease in other areas. Some, like Watson, considered scientists incapable of policing themselves since they would never agree to anything that would impede their work.[40]

Asilomar proved to be a most tumultuous event in scientific circles. It is impossible, in the space allotted here, to provide the reader with the proper level of intrigue and suspense that surrounded the moratorium and the conference itself. Looking back, many young scientists wonder what all the fuss was about. At the time, however, many were concerned (even alarmed) about the biohazards and research that could have potentially deadly results. Today, most scientists shrug off the concerns as alarmist since we know so much more and understand the process so much better. One hopes they are right.

HYPE OR HARM?

Were the NIH guidelines needed, or was this mere smoke and mirrors from nerds who were caught up in the spirit of Vietnam protest and so wanted something they could protest themselves? Certainly that had something to do with it, but precedent had been set. In addition to Berg's tinkering with the potentially deadly SV40, the cases of Stanfield Rogers and Martin Cline figured largely in this story.[41] Although W. French Anderson, who worked with colleagues at NIH in the early seventies, is credited as being the "Father of Gene Therapy," Rogers and Cline could easily be seen as its premature undertakers. (Anderson ran into his own scandalous difficulties when he was charged with child molestation in August, 2004.) Rogers pushed the envelope too far and too quickly by practicing gene therapy on a live, young child with an incurable disease. Rogers (and 10 years later, Cline, for similar reasons) was censured by his peers in 1971. His funding dried up almost overnight. He retuned to experimenting on plant viruses because the field of gene therapy had been closed to him. The reason this case deserves mention is that much of what Rogers did is practiced today. Even as they censured him, many scientists said that what he had done would eventually take a center stage. And it did.

The upshot of the Asilomar Conference turned out to be the NIH guidelines, but those guidelines were not easily forthcoming.[42] The first draft curtailed many experiments being done at the time but future drafts permitted them as scientists understood more, or at least thought they did. The last thing most scientists wanted was a ban on any area of genetic research that might prove enlightening or beneficial, even if early results proved dissatisfying or even possibly dangerous.[43] But their own fears were palpably tested by another fear: the reaction of the public and the misinformation of the media. Scientists called for the aggressive tack taken by the NIH because they did not want "housewives looking over their shoulders" in the labs.[44]

While the media and nonscientist writers (such as myself) were to blame, some of the blame rested on the shoulders of scientists.[45] In one celebrated case report, a highly respected science reporter, Janet L. Hopson, wrote a piece for *Smithsonian Magazine* after spending more than 90 days in a recombinant DNA lab.[46] She wrote about her experiences, including the nearly complete disdain for the NIH guidelines shown by the scientists working in the lab. Moreover, many of those scientists discounted the chance of anything hazardous in the least. And the head of the lab had even been a signatory of the Berg recommendations that had initiated the NIH guidelines in the first place.

The guidelines, however, do not have the force of law (as Pollack wanted). In conversations with some scientists, I have found that some are completely satisfied while others have a cavalier attitude about the guidelines.[47] In many cases, the scientists argue that laypersons simply do not understand the boundaries, the technology, or the manner in which the tests are conducted. This is unquestionably true. They are also quick to point out that no one in his or her right mind would do anything to jeopardize funding, so they comply with the spirit, if not always the letter, of the law of the NIH guidelines. When I asked about those labs where potentially dangerous experiments were underway, most said that precautions were taken, but not overly strenuously. Is the gain greater than the potential risk?

UNPREDICTABLE YOU

The Berg letter and subsequent recommendations argued that very little could be predicted about the behavior of DNA. For example, Berg had argued that DNA was unpredictable with respect to its expressing itself in the host, and whether these results (rDNA sequences) would be hazardous to humans, or whether the new DNA sequence would produce the desired (or undesired) elements in the new host.[48] What has continued to drive the controversy has been a near-complete volte-face by most scientists on this topic. In the early 1970s, many agreed with Berg and the recommendations, though the level of their agreement could certainly be said to vary greatly among scientists. At any rate, most agreed with the resulting guidelines. Today, most scientists would argue that the fears were unjustified or have since been proven to be less serious than once thought. There remain, however, many unanswered questions about DNA, especially its transgenic applications in plants and animals (with potential human implications) that scientists remain equally divided over. Because such a high level of scientific expertise is required to understand the debate fully, laypersons are likely to be left out of the debate entirely or, if allowed into the debate, will be shunned by scientists who fear they will hyperbolize both the dangers and the risks and thus threaten true research.

GUIDING THE GUIDELINES: REVISING THE REVISIONS

The Falkow memo of May, 1975 and the subsequent Falmouth Conference of June, 1977 helped begin the guidelines' revisions, especially with respect to *E. coli* and how it should be used in various rDNA experiments.[49] Two other revisions came in 1979, leaving us with what remains, by and large, the guidelines followed today. The guidelines that grew out of the Asilomar Conference are substantially altered in their form that appeared only four years later. Clearly there was some disturbing evidence brought by the Falkow memorandum on *E. coli* and some from the Falmouth Conference.

What has heightened concern in recent years is that the evidence for the relaxation was not based on hard scientific data but on the plausibility of the evidence to that point. Some have argued that the pressure from scientists to relax the guidelines created the changes, not what was understood in 1975 or any additional information that would have called for the more relaxed 1979 revisions. Indeed, as federal legislation moved forward (most notably abetted by Senator Edward Kennedy in 1975), pressure came to bear from as far away as Great Britain. The NIH (as well as Senator Kennedy's office) was flooded with mail (most all from the scientific community) opposing the guidelines as too restrictive.[50] Furthermore, it opposed any ban on any kind of research. As Krimsky points out, the battle was not only for freedom of scientific discovery but also for huge financial investment potential (as the founding of companies like Genentech in San Francisco in 1976 would later show).[51] "The battle," writes Krimsky, "was won by the scientists, for better or for worse."[52] Given the inherent conflict of interest, we can say it is not an unqualified good.

GENE THERAPY

Following 1979, what came to be known as gene therapy began in earnest (or continued) in nearly every location where minimal facilities were found.[53] The promise of disease control or eradication via gene therapy (swapping out bad genes, introducing good ones, or both) created bounteous levels of funding.

Initial research did not disappoint as the potential for halting, retarding, or completely controlling diseases like hemophilia, sickle cell anemia, cystic fibrosis, and emphysema rose with each new discovery.[54]

Gene therapy held such promise because it followed a logical progression: isolation of the abnormal gene from a normal patient; removal of defective cells (say, in a patient with sickle cell anemia); insertion of the normal cells; and reintroduction into the patient's body. But then this very process also raises fears concerning gene expression, precision, and the loss of human genes.[55] The discussion about recombinant DNA and its possible dangers had all but ceased.[56] The debate did not cease—and still has not—but the revised guidelines did much to bring closure to that part of the discussion. Not unlike President Lyndon Johnson's War on Poverty, President Richard Nixon's War on Cancer did not succeed or, at least, in a manner that was originally expected. Many advances were made, but by the end of the decade cancer research, while certainly further along that it had been, did not advance nearly as far as was originally hoped. Lack of funding could not be the cause. Perhaps it was the science?

SO MUCH HOPE, SO LITTLE HELP

While much hope rests on gene therapy, there have been few positive results to date. Why? One reason is the nature of DNA research and technology. The promise of somatic cell gene therapy (correcting gene defects in the body cells of a patient), for example, proved unlikely to spread beyond a narrow range of monogenic (one-cause) hereditary defects.[57] The reasons are both simple and complex. Predicting how DNA will work in a new host is only marginally more accurate than predicting how dice will roll at a crap table in Las Vegas. Another reason is the nature of various human diseases themselves. For example, gene therapy was at one time thought to hold great promise for treating high blood pressure. The reasoning was not without merit. Find the gene that causes it, cut it out, splice in one that does not, and blood pressure returns to normal. It has now come to be understood that diseases like high blood pressure are multifactorial; there is more than one reason why a person has it. There may be an inherited predisposition to a particular disease but other factors such as weight, improper diet, lack of exercise, and so forth are equally if not more important. Changing the gene will not change behavior.

OBSTACLES TO SUCCESS

The process is not so simple, either. Many obstacles still remain in spite of all the considerable progress that has been made. As one scientist put it,

"Therapies are brand-new. There are no precedents for them."[58] In practical terms, this means that whatever a scientist may do to a lab animal—say, in a rat or a mouse—cannot simply be transferred to a human and be expected to work. In fact, early evidence points to a disturbing finding: what has aided mice in the control of diabetes, for example, may in fact cause a deadly form of cancer in humans, all the while aiding the diabetes.[59] It brings to mind Francis Bacon's (1561–1626) quip that patients often were cured of diseases only to die from the cures.

Other diseases simply do not respond as scientists think they should.[60] Following the establishment of the NIH guidelines, scientists began to see how genes worked in certain sequences. Of course, for scores of years— certainly at least since Mendel and through observation before that—scientists knew that certain traits were handed down and certain genes were responsible for those traits. What they did not know were which genes were responsible for which trait. In the 1990s this led to a great push to identify all human genes and what they do; this is commonly referred to as mapping the human genome (a more detailed discussion follows in Chapter 6.)

At first it was thought that it would take many, many years to map all 100,000 to 150,000 human genes. After about a decade of research, however, scientists discovered that there were not 150,000 or even 100,000 genes, but only about 25,000, and that the map could be completed much sooner than expected. As I write these words at the beginning of 2005, nearly all scientists believe this process will be completed this year. So, once we know what each gene does and where it resides on the human genome map, can't we cure every disease?

It would seem so, but few scientists would agree. Think back to Chapter 2 and those sequences that were discussed. Very powerful, very complex computers have mapped these sequences so that nearly anyone can identify a sequence of As, Gs, Cs, and Ts and determine where it resides and what it does by looking it up on the Internet. But among all those codes are other sequences that some scientists refer to as junk DNA. It is called this because no one knows what it does or why it appears. Some scientists object to the junk designation and it appears there's good reason for thinking so. Only 50 years ago, any scientist looking at just about any sequences would have thought them junk because they didn't know their purpose. Perhaps, the current argument goes, we shouldn't be so quick to call something junk just because we haven't identified it yet.

GENE THERAPY: TWO STEPS FORWARD, SIX BACK

Then in 1999 there was the celebrated case of Jesse Gelsinger that resulted in the complete cessation of at least one kind of gene therapy.[61] Gelsinger had an unusually mild form of a rare and often fatal liver disease (ornithine trans-carbamylase, mercifully OTC) that required a complicated diet. The disease prevents or obstructs the removal of ammonia from the body. His younger brother had just died from the disease and a new gene therapy protocol held out promise to him, and others, too. There was one small slip-up: the cure killed him in less than four days. What is more, it is not certain whether this was the first death from gene therapy or only the first that we know about. The story is tragic, filled with some mystery and even suspicion as researchers may have cut corners owing to pressure to get a cure to market. What could have been more compelling than helping a young man who wanted to save himself and others, if only it had worked?

TOTIPOTENT VERSUS PLURIPOTENT

Research into gene and stem cell therapies has slowed because of the process itself. Look back at the four steps outlined above. One of those steps includes establishing libraries of genes (genomic libraries) To do that, scientists must have cells to work with and these cells come from somewhere. Scientists have labeled the two types of cells as pluripotent and totipotent, and herein resides much of the controversy. Speaking broadly, totipotent cells are those that have not been given a genetic signal or command to grow into a certain kind of tissue—heart, stomach, kidney, or the like—just yet. They are undifferentiated because they can replicate themselves or become something else entirely.[62] Pluripotent cells are cells that have already gotten the command signal and begun the process to become a specific tissue and so have fewer options. That is, they are already becoming heart tissue or stomach lining cells. Pluripotent cells can come from any adult, while totipotent cells must come from embryonic cells—in other words, from fetuses.[63] If you think this raises the abortion debate all over again but from a slightly different angle, you'd be right.

Some scientists believe that taking embryos from fertility clinics, where they are scheduled for elimination anyway, should be an option for research. Some think securing them from umbilical cords is a safe and uncontroversial means of harvesting them. Some doubt the quality of umbilical cord harvests. Others worry that either approach is too much like the abortion process and will lead some women—given the large amounts of money made available for this kind of research—to become so-called hatcheries (reminiscent of the hatcheries in *Brave New World*) for this purpose alone.[64]

But there is more to it than even this. Scientists still need to know more about the way in which cells specialize or know how to become the various tissues of the body. In the earlier discussion of the revisions for the NIH guidelines, it was pointed out that not much had changed in the knowledge base that led to those revisions from 1975 to 1979. Scientists still do not know why one cell becomes heart tissue and another becomes stomach lining tissue or goes into making a kidney. What happens when cells meant to repair the heart suddenly and inexplicably become liver cells or, worse, mutate into some form of cancer? As one scientist has put it, "Although genetic engineers can cut and splice DNA molecules with base-pair precision in the test tube, when an altered DNA molecule is introduced into the genome of a living organism, the full range of its effects on the functioning of that organism cannot be known." [65]

One reason for this is what McClintock found out about jumping genes (discussed in Chapter 2). The Human Genome Project has discovered that less than 2 percent of the 3 billion bases in the human genome encode the genes. About 50 percent is made up of genetic elements that originated from other, nonhuman sources, one type of which moves, or jumps, from place to place in the gene.[66] As yet, there is no way to track these nonhuman elements.

Then there is the problem of rejection.[67] Great strides have been made in organ transplants but not without numerous drugs to reduce the possibility of organ rejection. Rejection also looms large in the placement of cells from one host to another, and we are many years away from developing drugs to address this different rejection problem. Finally, we have now discovered that some organisms are, to a greater or lesser degree, simply resistant to any cloning whatsoever.[68]

THERE'S GOLD IN THEM THAR CLONES!

This lack of knowledge has done nothing to slow the growth of genetic industries. The founding of Genentech in 1976 (mentioned earlier) began the exploitation of hormone research in the San Francisco area. It proved wildly successful early on by amassing over half a billion dollars in financial worth before ever producing a single product.[69] This was followed by Biogen, a company that focused on interferon for cancer research; it was founded by Charles Weissman (University of Zürich) and Walter Gilbert (Harvard University).[70] The two sold 16 percent of the company in 1979 to Schering-Plough and the resulting Biotech Co., Ltd. became the first DNA-oriented company to link itself with a multinational drug maker and distributor. This was followed in the 1980s by what some have referred to as a gold rush of companies, not unlike the establishment of Silicon Valley in the computer industry. Other companies such as Cetus, Genex, the Genetics Institute, the Collaborative Research of Waltham, Massachusetts, and Bethesda Research Laboratories in Washington, DC have since sprung up. By the mid-1980s there were more than 150 small companies in the United States alone working in genetic engineering and/or advanced biotechnology.[71] All of these companies can claim some link to either Nobel laureates (such as David Baltimore, whose early work with Berg proved pivotal in the establishment of the NIH guidelines) or other geneticist luminaries.

THE ETHICAL DEBATES

On one hand, this has led to the rapid growth in this field and done much to push forward the research in an important area. On the other, it has raised serious ethical questions about the link between academic work and research and work guided by the larger companies (and shareholders) who own them. Some have raised the question of whether these scientists can pursue research unimpeded by the pressure to succeed, both in their universities and in their companies.

Certainly no one other event in genetic engineering created as much controversy while opening up new avenues of research as what occurred on July 25, 1978 in an English hospital. Louise Brown, weighing in at 5¾ pounds, became the world's first test-tube baby.[72] Although there were a few news reports of her twenty-fifth birthday, she has since been eclipsed by Dolly the sheep and CC the kitten (about which there will be more in Chapter 5). The story of Brown appeared in the press with much controversy and even some horror. But the *que sera, sera* attitude with which we now view her birth underscores that famous Alexander Pope line that, in time, eventually we "endure, then fondle, then embrace" even the greatest problems regardless of potential vice. While human cloning is still viewed by many with nothing short of horror (four out of five Americans see it as either "against God's will or "morally wrong"), it may be that time will cause many to see it as ho-hum science in the long run.[73] Even so, the debates about cloning have raised serious ethical questions that remain unresolved.

Much confusion still exists regarding what the federal government is doing about stem cell research and gene therapy in general. During the 2004 presidential election, some tried to make political hay by pitting President George W. Bush against contender John Kerry. But the Bush position was actually broader than that of Bush's predecessor, Bill Clinton.[74] Under President Clinton, research was allowed but any lab receiving federal funding was barred from using that money to pursue stem cell research. Under Bush, funds may be used for stem cell research but with the restriction that the lines have to be already in use, not newly established lines.[75]

The discussion about genetic engineering has not been limited to science alone, either. A spate of books that spill over into philosophy and theology have been arriving for nearly two decades now in an effort to at least identify the ethical issues and the religious quandaries raised by genetic manipulation.[76] The debate is not an easy one. On one hand, most would agree that no one wants to clone humans, not only because it seems so odd and surreal but also because we do not want the risk of having it available for billionaires alone, or risking a copy machine-like production of 100 Hitlers. Suddenly, we are faced with seriously ethical questions that stagger the mind with real-life scenarios only seen in films such as *Multiplicity* and *Blade Runner*.[77] On the other hand, just how easy would it be to deny young parents who have just lost their only child a chance to recover him or her through the cloning process? If all of this was not enough, throw a few lawyers in the midst and the discussion suddenly rises to new heights (or depths, depending on one's view).[78]

According to one critic, high-tech science has spread as many genetic weaknesses and deformities via neonatal surgery as it claims to have helped. Social policy has combined with neonatal surgical successes to make offspring of what could be called pathetic creatures likely.[79] Others are horrified by advocacy of intervention in childbearing decisions or other decisions that are the outgrowth of genetic research.[80]

MORE QUESTIONS THAN ANSWERS

Is there more? Much more. What has been covered here is just the tip of the proverbial iceberg.[81] For example, nothing has been resolved regarding insurance reportage after someone has undergone gene therapy or has had gene screening for a particular disease.[82] Some have argued that employers have a right to know if a certain employee has a gene for an incurable disease, whether that employee has it himself or will produce offspring who will have it and will drag upon the company's insurance claims. It doesn't help that the precision of the screening is still much in debate.

Some incredibly interesting work has been done in other areas, as, for example, in using DNA to reflect upon the health of historical figures or to catch criminals (see Chapter 8 for more on the latter).[83] We cannot discount the value of knowing certain markers for sickle cell anemia, thalassameia, or breast cancer (BCRA-1). Very positive research is being done in an effort to "teach the body to kill cancer."[84] Is it possible that there could be any serious argument against anything that sounds this wonderful?

Yes, many, in fact. Will genetic manipulation be available to everyone or only to the well-to-do and well-favored?[85] While the advent of gene therapy involving somatic cells is debatable, what appears not to be arguable is germ cell therapy (or the correcting of a genetic defect in the germ, or reproductive, cells of a patient to prevent the same defect in offspring),[86] Since germ cells in the human genome constitute tissue independent from the rest of the body, many feel that tinkering with these is not acceptable, at least in humans.[87] Germ cell therapy is ongoing in plants and animals, but fears about the push for manipulation of germ cells in humans are palpable. For example, the so-called designer babies point to this distinct possibility, for if genes such as for intelligence and good looks can be spliced into genes, will succeeding generations be able to pass up the chance to design perfect children? Others are concerned that we may well be splicing out certain altruistic feelings for the downtrodden and the unfortunate. Finally, commercialization of gene therapy has drawn out many and varied critics.[88]

Are human clones possible? It seems likely, if one is to believe the Italian physician Severino Antinori who announced in 2001 that he would create the first human clone.[89] He is not alone. Before Antinori came physicist Richard Seed promising to clone human beings. Seed, knowing that U.S. law would likely prohibit him, promised that if he could not do it here he would go to a country that would. His announcement sounded many alarms and even sparked a bill in Congress (it never passed). President Clinton's Secretary of Heath and Human Services, Donna Shalala, called Seed a "mad scientist."[90]

HUMAN CLONING: TOO MANY MINI-MES?

Will human cloning result, as some have argued, in human clone farms of headless mice and men?[91] Some may scoff that this nonsense is the hysteria of doomsayers. Scientific know-how is, however, making such scenarios less nonsensical and more probable with each new discovery. Of course, one should not go overboard, either, and attribute to cloning what cloning can never do. If a child should die, for example, could parents save his or her DNA and resurrect the dead? No, or yes with a qualifier. The resulting clone may not look exactly like the deceased and certainly any change in environment would change his or her behavior and personality. Even identical twins, regardless of how long they are dressed alike, develop different personalities. Rumors, as yet unproven, indicate that the wealthy have funded private companies to freeze DNA for their own use as needed.[92] Still, even a remote chance that someone would want to do this staggers the mind and disturbs the will of some. It does not help that some writers try to reassure us that humans will continue to maintain society "for some time," as if to say, "not forever."[93]

Cloning, at least in the plant kingdom, is alive and, well, blooming.[94] Genetically modified seeds both for flowers and edible plants continue apace, but not without controversy, as we shall see in the next chapter.

Research is not likely to slow, either, with the 2004 passage of California's $3 billion ($300 million a year) initiative for embryonic stem cell research. This has led not only to great interest but also to panic in other states where geneticists now work.[95] The fear is that California will get not only all the best, but all researchers interested in this kind of research. Whether this proves true or not remains to be seen. What it does mean for now is that the issue will not go away until the funding does. Where will it all end? If we can clone humans, why not clone great geniuses, the super-healthy to create a super race, or genetically identical humans for study, or control for the gender of children, or stock organs for future use?[96] These are issues that continue to surface from time to time with just about equal support both for and against.

Money for funding genetic engineering is not the only issue, either. Some fear money could drive individuals as well. Suppose a woman could sell a fetus for $10,000 or even $20,000? Assuming that these unborn would only be taken to half-term, this would supplement some incomes generously. Would this not wrongly encourage some to pursue this dreadful avenue? Or, suppose it isn't this gruesome scenario but another, cloning for the sake of producing cells, tissue, or even organs?[97] While one hopes these turns of events will never come to pass, the advent of cloning has made them possible. Sadly, our human history indicates that we have often taken the low road when a higher one is available. For some, the whole debate smacks too much of Josef Mengele, the concentration camp doctor and mad scientist who during Hitler's regime caused untold pain and suffering with his so-called experiments.[98]

Finally, none of this has addressed concerns that are uppermost in many minds after the September 11, 2001 terrorist attacks: biochemical warfare. Is it possible for terrorists to use some of this technology for the destruction of the planet? The answer is yes, it most definitely is, should the technology fall into the hands of the wrong people.[99]

In many ways, the fears, the predictions, and the promises that originated at the Asilomar Conference have all come true.[100] We have learned much but we have also saddled our culture with a great deal more. As is often the case with scientific discovery, the technology has surpassed the ethics, perhaps even bypassed it. Many people, such as the poet Theodore Roszak, contend that science and its blind lust for knowing has corrosive effects on culture. Others, such as Leon Kass have argued that "We have paid some high prices for the technological conquest of nature, but none perhaps so high as the intellectual and spiritual costs of seeing nature as mere material for our manipulation, exploitation, and transformation ... [I]f we come to see ourselves as meat, then meat we will become."[101] The debate is today evenly divided between those who believe we have become that meat and those who still believe we have the power to be masters over it.

4

Frankenfood or Miracle-Gro? Agricultural Applications of Genetic Engineering

With the advent of genetic engineering since 1971, geneticists have made great breakthroughs with plants and their reproduction. By modifying certain foods (usually referred to as genetically modified foods, GMs), scientists are able to speed up, slow down, or make otherwise resistant to various diseases or certain temperatures various kinds of foods, including beer, wine, bread, yogurt, and cheeses, to name but a very, very few.[1] While beer and wine have, of course, been around for hundreds of years, the way in which we manipulate them today, along with many others kinds of foods and even seeds, provides for greater food production, better use of arable land, cultivation of lands that may never have been used for farming before, and more plentiful (both in number and variety) foods for the world's consumption. It has also been, like every other procedure so far examined in this book, a source of controversy. It doesn't help matters much that science fiction is becoming more science than fiction. Even the amber-trapped creatures-turned-dinosaurs idea (as was seen in the *Jurassic Park* movies) has more credibility than not.[2] That should not be too surprising, given the way some people regard the ability of scientists. According to one CEO of a genetically modified food co-op, "Of course it will work. Give a scientist enough time and money and he can do anything."[3] (The question of whether he should or not is tabled for now.)

Part of the controversy has more to do with the fact that we change food—even the very structure of the plants—than it has to do with anything else. People will tolerate many things but most will not tolerate tinkering with their food. One researcher has argued that "The genetic engineering of our food is the most radical transformation in our diet since the invention of agriculture 10,000 years ago."[4] Recall that only 15 years ago we had a national hand-wringing over a substance called alar added to apples. Alar allowed apples to ripen more slowly, retain their color longer, and provide for fewer rotten apples coming to market. However, because alar was also identified as a possible carcinogen (cancer-causing agent) in food, everyone from John Q. Public to Meryl Streep made a federal case (literally) about it. In order to match the amount of alar that causes cancer in mice in a human, that individual would have to eat 750 alar-treated apples every day for 70 years.

Alar, one of the more innocuous substances in this controversy, provides a nice backdrop to this discussion because it highlights both the need for concern and the need for perspective. Of course, much of the debate centers on what constitutes a so-called right perspective. If your view is that any amount of cancer-causing substances (or any other toxic agents) in foods is wrong, then much of the debate is over and you should not eat another bite of food ever again. If your view is that no amount is too much as long as it protects jobs, protects farmers, and makes a profit, then likewise the debate is over and you need read no further. On the other hand, if you haven't quite decided what is right, perhaps this chapter will shed some light on the subject.

To be sure, people have been tinkering with food for thousands of years.[5] Farmers have rotated crops and taken shoots from one plant and grafted them onto another, hoping for a stronger, more fructifying plant. Nothing could be farther from controversy than this. But where there is much controversy is the point at which scientists began changing the internal structure of foods—plants and seeds—and putting them in the hands of farmers to use. Although the so-called new biotechnology is hardly adolescent, like adolescents it has already sparked heated debate over injecting genetically engineered hormones into dairy cows to increase their milk production. This caused some American grocery stores to ban milk from those cows, and created controversy over genetically modified seeds, but it produced huge profits for giant food cooperatives. It has also been a lifesaver to farmers who spend millions of dollars on chemicals trying to ward off diseases or brought million-dollar lawsuits because of crops lost to insects. Such techniques (in the latter case, a synthetic DNA probe) have even been responsible in detecting shellfish-related diseases when nothing else would.[6] It has also given new life to organic farming, a popular pastime among 1960s hippies that's making a strong comeback among today's yuppies and Gen-Xers.[7]

Anyone who objects to helping farmers, decreasing the insect population, and curbing diseases (all the while creating higher agricultural yields) would have to be crazy, right? Not really. The trouble comes when you pull your chair up to the dinner table and suddenly that steak, potato, pumpkin bread, turnip, and sliced tomato look like something from outer space, at least in your mind's eye. That's when people begin to wonder, and rightly so. On the one hand, if we are what we eat, how is it possible that when we eat these genetically modified foods nothing happens other than nutrition? On the other, if it is so bad, why can't we simply stop it?

The answers are not altogether easy. In the first place, the new biotechnology industry represents a heretofore unseen concentration of corporate power and wealth in two fundamentally important areas: health and food.[8] As more and more powerful conglomerates merge, the ability of one or two to control the production of all food is immense. This is not altogether a bad thing if such conglomerates can be trusted to do the right thing. That, of course, is the real "rub," as Shakespeare might say. By 1999, for example, five companies (AstraZeneca, DuPont, Monsanto, Novartis, and Aventis) held 60 percent of the pesticide market globally, a quarter of the commercial seed market, and about 90 percent of all genetically modified seeds.[9] While such a concentration makes it easy for farmers to gain access to seeds because it is easier to distribute seeds globally, it also presents a problem when trying to tell one or more to quit doing something or to begin to do something differently. There simply is not much leverage if all five of those companies agree, even if there has been strong opposition to their genetically modified processes as there has been over the last 10 years (especially in the European Union, see below). Moreover, when all five can point to tremendous successes in Third World crop production and arable land improvement, it makes any debate seem almost moot.

Almost moot. While these and other companies have fought strongly for GMOs and other food-tinkering stuffs with their unquestioned successes, it has done little to change the debate or its intensity from opponents. In some cases, it could be said to have heightened it. It's easy to see why. From mad cow disease to dioxin-laden animal feed and hormone-fed beef, we humans simply do not like having our food tinkered with, even if you promise to make it better.[10]

In many ways, we stand at the gates of a new Eden, ready to create, via genetic manipulation, perhaps not so much a new heaven and a new Earth, but certainly that which goes into making up a new Earth. This is the great promise of the new Eden; it is also the great peril.[11] For example, injecting the growth hormone rBST into cows heartily raises milk production, as noted above. This is good news for areas where milk is scarce and demand is high. It would allow U.S. farmers to produce more milk for even wider distribution, thus increasing their sales, improving nutrition, relieving death from poverty and malnutrition, and reducing overhead costs for dairy farmers. Unfortunately, science does not provide an answer about the possible side effects of rBST, as studies are inconclusive as to whether it adversely affects the human immune system.[12] On the other hand, there is *Baccilus thuringiensis (Bt),* a bacterium in the genes of about 33 percent of the nation's corn. Corn genes were modified to allow the release of this bacterium that acts as a built-in pesticide to certain insects. The prevalence of corn is also one of the reasons why all of us have doubtless eaten genetically modified foods. Because corn could not be successfully grown in the South without having *Bt* to combat pests, *Bt* has been a target of genetic modifiers for years.[13] It has also been linked to the reduction in the number of monarch butterflies via rBST-infected pollen, or not, depending on the research one reads.[14] Furthermore, while some large farming corporations have increased profits significantly, others have gone bankrupt; the yield of genetically modified seeds has not lived up to claims; the advent of patenting genetically modifying procedures has closed off research; and the claims of scores of new seeds from hybrids may mean wider variety, or less.[15] In other words, for every good claim there is a downside, for every step forward there appears at least one back, and for every new claim there is an opposite counterclaim.

Another fundamentally important issue is the ethics of genetic modification in general. While it may be easy to see the merit of such claims in the case of animals (to be discussed in Chapter 5), many people are raising them now with respect to plants and seeds.[16] As we shall see, the response from America is quite different from that being seen in Europe, Asia, Canada, South America, Australia, and New Zealand.[17] In some cases, the very countries raising these questions stand to benefit the most from the positive claims made on behalf of genetically modified crops. In these countries and others, the issue comes down to whether using recombinant DNA produces that which would otherwise never be found in nature, or whether whatever results has any potential to create more harm than good. These are not easy questions because reputable scientists have come down on both sides of the question. Unlike the alar-treated apple case cited above, the evidence can be easily treated either to marshal a case for genetically modified foods or to restrict (if not prohibit) them.

Moreover, it isn't so much the artificiality of the process that makes it dangerous in the minds of most people (though this is, to be sure, some cause for alarm). Rather, it is not knowing what happens afterwards. Molecular geneticist Michael Antoniou puts it this way: "[What makes genetic modification seem dangerous] is the imprecise way in which genes are combined and the unpredictability in how the foreign gene will behave in its new host that results in uncertainty."[18] Antoniou goes on to point out that this inability to predict far outweighs the benefit of the intended change.

Some argue that the matter should be left to consumer choice. If, for example, I choose to eat alar-treated apples, so what? That should be my decision, the same as if I might choose to buy a leather belt instead of a synthetic one or buy silk as opposed to a synthetic derivative, shouldn't it? But the argument of consumer choice begs the question. As we shall see, consumers do not always know and the FDA (U.S. Food and Drug Administration) does not always provide enough information or the proper safeguards to make a wise customer choice possible.[19]

Would it surprise you to know that you have eaten genetically altered food? You have if you've eaten food purchased at any one of a number of well-known U.S. grocery store chains. As much as 70 percent of food in American grocery stores has been genetically modified, according to experts.[20] One researcher stated, "To even suggest that there may be ethical issues to be con-

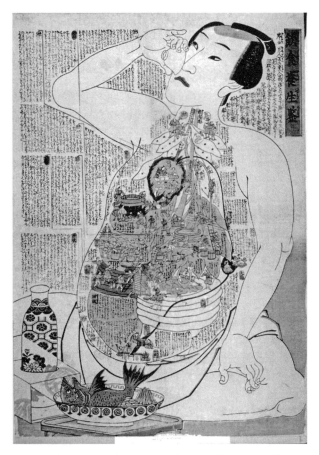

You *are* what you eat! Inshoky yojo kagami, Karai. From the National Library of Medicine, History of Medicine Collection.

sidered concerning the so-called technology itself may be a cultural leap for those immersed in the culture that has produced genetic engineering."[21] In other words, we may have inadvertently put the fox in charge of the hen house. This is because, as was pointed out in Chapter 3, genes (whether plant, animal, or human) are an unknown commodity when they are inserted into the same or different species. It raises, among other ethical questions the problem of ownership. Perhaps in the case of humans we might be quick to say they belong to that human. But what about plant genes? Who owns them? And what about animal genes? Are they owned by all of us or by the scientists who created them? Even in the case of human genes, the question does not yield to an easy answer. If a gene is taken from person A and combined with another gene in person B to create a third gene, gene X, who really owns it? Since it would not have produced itself in the natural course of DNA zipping and unzipping, are donors A and B really owners or merely vehicles?

DIAMOND V. CHAKRABARTY: WHOSE LIFE IS IT?

This leads to another important question, or at least one that plant genetic modification brought to us: is life—any form of it—patentable? It's not an idle question and it was posed long before now (in the 1920s, to be exact) by Luther Burbank, who argued that a man could build a better mousetrap but not benefit from plant breeding as a life's work because Congress had not passed a law enabling him to do so.[22]

At first blush, patenting life forms would seem the silliest thing on earth and that is certainly the way most scientists saw it. That is, until 1972 and a man named Anadana Chakrabarty, who worked as a biochemist at General Electric.[23] He had devised a slick little bioengineered phenomenon, a bacterium that would consume oil slicks; the bacterium itself as he devised it did not occur naturally.[24] What could possibly be better? Suppose another huge oil spill like the 1989 Valdez disaster occurs? Scientists could grab a handful of petri dishes, head to the site, and within hours this bacterium and all his fellow bacteria could be eating away at the slick. Almost before it starts, it's over. Well, okay, it isn't that neat but you get the picture. Then Chakrabarty did something no one else had. He applied for patent on the living, bioengineered, genetically modified organism.

American patent law has been around for a long time (since 1793) and was penned by none other than Thomas Jefferson himself. Patents can be obtained for "any new and useful art, machine, manufacture, or composition of matter, or any new and useful improvement thereof." Only one word has changed since 1793. Jefferson's word "process" became "art" in 1952. Chakrabarty claimed that both that little word and Jefferson's phrase "or any new and useful improvement thereof" allowed him to patent his bacterium.[25]

Still, most dismissed Chakrabarty and his claim as so much shooting blanks. After all, for more than 100 years at the time of the Chakrabarty filing, patent law had prohibited similar things, such as a patent for a fiber in pine needles. The court had rejected that claim, arguing that the composition of pine trees was not patentable any more than someone trying to patent a gem he had found and all others like it. This ruling came to be known as the "product-of nature" dogma and allowed for a distinction between man-made items and nature-made ones that had stood for more than 100 years until Chakrabarty. The Chakrabarty argument seemed simple: here an organism already found in nature had been tinkered with and a new, more aggressive one had developed that would consume oil slicks. Besides, Chakrabarty claimed a patent for a process, not really for life or a life form.[26]

But Congress, as is well known, seldom leaves well enough alone. In 1930, Congress passed the Plant Patent Act. Essentially, this said that plants could be patented if they were produced asexually. Chakrabarty applied and the U.S. Patent and Trademark Office denied him, basing its decision on the product-of-nature dogma. Chakrabarty appealed through the courts and his case reached the Supreme Court as *Diamond v. Chakrabarty* (Sidney Diamond was the then-commissioner of patents).

Nothing happens in a vacuum, and patent law (which remains a foreign language to those who have not studied it) is not immune to outside forces. The court already knew that this process relied on recombinant DNA developed by Boyer and Cohen. Recall from Chapter 3 that Boyer had combined forces with venture capitalist Robert A. Swanson to form Genentech in 1976. What Boyer had developed was not an organism so much as a process for moving DNA about in sexually incompatible organisms.[27] Genentech, as was pointed out, had hit the front pages of many newspapers with its hope for curing millions of people with diseases via DNA improvements. The government already had its hand in this work by way of the NIH and its sponsorship of various university outreaches in the genetic modification enterprise. Billions—not millions—of dollars hung on this decision and the court knew it. Moreover, nearly a dozen amicus curiae briefs were filed, most in Chakrabarty's favor and all by companies who stood to gain. Only one stood on the side of Diamond and the Patent Office. It had been filed by the People's Business Commission, headed by a well-known and much disliked social (some would have claimed, socialist) activist, Jeremy Rifkin. Rifkin's group claimed that this would be only the tip of the proverbial iceberg. Soon would follow animal patents and, of course, human ones. Does this ring a *Brave New World* bell?

Needless to say, Genentech and its followers laughed at the proposition that humans would be next. Of course, it hedged its bets. Some might make claims on procedures that could be used on humans. Other amicus briefs made the claim (correctly) that too many plants and animals to count had already been treated as property. Giving Chakrabarty a patent would do nothing more than approve what had already been done. Of course, no one wanted to dwell too carefully on this because only Chakrabarty had applied for a patent on a process of creating new life (even if Chakrabarty had not relied entirely on recombinant DNA). Chakrabarty's case survived court in 1980 and the effect made it "a composition of matter or manufacture," as Jefferson had innocently penned.[28] But the case survived only barely; it was a 5 to 4 decision that a live, human-made microorganism could be patented.[29] The once-denied patent was granted after all.

In the aftermath, Congress added to its meddling with the Plant Variety Protection Act (PVPA) in 1970. PVPA "provides patent-like protection to new, distinct uniform and stable variety of plants that reproduce sexually." The PVPA was revised in 1994 and extended this protection for 20 more years, just like a patent but not administered by the U.S. Patent Office.[30] Other cases have added to or muddled the matter. The Bayh-Dole Act (a Patent and Trademark Laws amendment), passed in 1980, made it much easier for university scientists to perform patent work while receiving government money. In 1986 Congress made it easier still with the Technology Transfer Act for government scientists to patent their discoveries. Only a patent infringement case (*Regents, University of California v. Genentech, Inc.,* 1999) caused even a faint slowdown in the biotechnical, genetically engineered onslaught.[31] The case was a boon to the biotechnology industry; it remains to be seen whether it will become a boondoggle for everyone else. Why a boondoggle? Robert Shapiro, CEO of Monsanto Corporation, may have summed it up best when he said, "We are learning about biology at a level and at a rate that is absolutely unprecedented in human history. There is an enormous potential space to be filled, and the stakes are very high. We want to be able to occupy the most valuable territory."[32] Even if we assume the best of motives of all included, is it wise to invest this much power in the hands of so few?

The Chakrabarty case is by far the most familiar case, but it is not the only case. Nor has it been resolved, even though the case was won. It raises a number of issues, not the least of which is intellectual property.[33] Can someone own something that nature produces but man rarefies or otherwise enhances, enlarges upon, or materially changes? It would seem so.

With plants, the question becomes all the more tricky. For years—make that centuries—humans have been tinkering with food to make it better. In the field of horticultural studies, hybrids have been a staple field of research since 1950.[34] Do we draw an imaginary line of demarcation between what would occur naturally and that which requires human interference? If so, a good deal of what we have been eating is off-limits. If we draw the line between some things that require human intervention and some that do not, what are the criteria? And what of the hybrids that occurred naturally, so to say, by blending the so-called wild ancestors with other plant varieties? Most scientists argue that the distance from wild ancestors to any plant even our great-grandfathers planted is as large as it is between them and the first man.[35] Or, do we isolate those areas where plant experimentation is very new from that which has been around for—five years? Ten? Decades? Fair game? If we choose the latter, what standards do we set and how do we differentiate between the techniques that are acceptable and those that are unacceptable, since much of what we now have might have begun as a scenario we might have dubbed unacceptable? Moreover, if we sideline one form of genetic modification, will we be able to allow all others? For example, the horseracing and floral industries have relied on some form of genetic tinkering or cross-breeding for decades.[36] What is different, of course, is the degree to which it can be done. A traditional cross-breeder could never breed a tree with, say, a carrot. But genetic modification now allows a gene from the tree to go directly into the carrot without dragging all other tree genes with it.[37] Furthermore, plants are much more "plastic" with their genes and chromosomes, allowing for an outsider to double or even quadruple the chromosome set.[38] What this means is that geneticists have much to work with and many avenues to explore. Again, these are not easy questions that yield to easy, black-and-white answers, only very subtle shades of gray.

Gray may not even begin to cover it, either. In September, 2000 the *New York Times* ran an article about a genetically engineered corn that contained the protein Cry9C. This would not have been news by itself. What was news, and what sent something of a tsunami through the genetic engineering world, was that while Cry9C had been approved for cattle and other animal feed, it had not been approved for human consumption. Yet, there it was, in a number of nationally sold brands of corn tacos.[39] Apparently it was now possible to have genetically engineered food in food that had not yet been approved for genetic modification.

Early research into recombinant DNA using plants revealed that efforts to introduce alien genes allowed for the DNA to enter the plant but not to replicate. Scientists learned that what had to be done to achieve successful replication and copying required sequences of DNA to be recognized by enzymes involved in the process.[40] This discovery allowed for a better understanding of what took place when these foreign bodies were added and provided insights in a number of directions, not only for plant viruses but also for other applications. While the end result has not necessarily been the so-called Frankenfood, it also has not been happily received by everyone.

EARLY BLUNDERS

Genetic manipulation in plants did not begin altogether smoothly. In Chapter 2 we stressed how unpredictable genetic insertion and/or cloning is. Once the gene is inside the new organism, geneticists know how it should work, but the genes don't always get the right instructions and so proceed on their own. While genetic activity is perhaps more predictable than a coin toss, a gene, like any living thing, cannot be told what, unswervingly, to do. When genes go awry, the end result is not the 50-foot-tall woman from science fiction movies but it can be something that no one thought would happen. Not only can plant follow-through not be predicted, but even more unpredictable is the public's response. In these examples of early blunders, both gene unpredictability and public response play major roles.

ICE-MINUS

The year 1984 has always been seen as ominous, mainly (or exclusively) because of George Orwell's *1984*, wherein Big Brother (a euphemism for what happened under communist totalitarianism) would take over and change everything, including history. Genetic engineers may purposely have waited a year, but in 1985 (and the year following) two recombinant DNA products were released into the environment in direct violation of federal rules on rDNA.[41] Two bacteria, *Pseudomonas syringae* and *Pseudomonas florescens,* were the culprits. The organisms, developed by the University of California, had been developed by taking away genes rather than adding them. The press had dubbed the new organisms "ice-minus" because a genetic company, Advanced Genetic Sciences (AGS), had planned on marketing them as Frostban. The ice in ice-minus referred to the gene, the minus to its removal.

A mutant of the common bacterium *Pseudomona* is found in potato plants.[42] *Pseudomonas* produces a protein on which ice crystals will form quickly. The idea of ice-minus (and it seemed a very good one) allowed these bacteria to collect on leaves as soon as the temperature reached, or dropped slightly below, 32°F. Ice-minus as a gene deletion would decrease the likelihood of frost forming, or forming as rapidly.[43] Given the marketing name, readers can guess that crops severely and adversely affected by the cold—strawberries, for example—would be immune, so to speak, to frost. Sprayed with Frostban, any crop adversely affected by frost would be protected. Suddenly, even International Falls, Minnesota, the freezer capital of the world, could grow strawberries. Okay, perhaps not that cold, but certainly areas where strawberries would not grow owing to prolonged frosts could at least consider this as a new crop. More strawberries could be produced and, because of supply, prices would come down.

AGS wanted too much, however. After getting permission to test it on strawberries, the company also sprayed it on trees on the roof of its corporate headquarters. Government permission was withdrawn and the company had to pay $13,000 in fines, the first such fine levied by the U.S. EPA. But why? Mainly because the use of ice-minus in an uncontrolled area evoked fears about two other not-quite-similar blunders using *Pseudomonas syringae*: Dutch elm disease and chestnut tree blight. These diseases were inadvertently caused by the release of organisms in ecological areas that did not have these diseases before the release. The result proved disastrous. Both the American elm and the American chestnut tree have all but been eliminated by diseases resulting from this release. In other words, the fears were ecological in nature. Would widespread spraying create an ecological imbalance?

Later studies argued that the result of the ice-minus release would be negligible, but the damage from the fear of ecological disaster had been done. That, combined with the Dutch elm disease and the chestnut tree blight blunders, proved too much for ice-minus. Even with a clean bill of health—which it never fully got—marketing ice-minus over and against ecological fears proved impossible. The case raised concern about uncontrolled testing, concern over widespread damage, and whether or not appropriate checks and balances were in place to account for the same. Clearly, scientists involved in such companies could not be relied upon to exercise good judgment (or so the argument went). American regulatory constraints appeared to be too relaxed, allowing for progress before all potential problems had been addressed.[44]

Others have argued, however, that the fears about ice-minus and the Flavr Savr Tomato (see below) were overblown. These fears, say proponents, have retarded growth in the industry and have led to countless delays in the discovery of medicines and even cures for deadly diseases such as AIDS.[45]

FLAVR SAVR TOMATO[46]

The idea behind the Flavr Savr tomato could not have been better. All of us grocery shoppers are always elated when spring arrives. Spring and summer mean produce—fresh, delicious, ripe produce. Fall rolls around and the produce declines. Then come the first snows and we find ourselves staring at canned or frozen goods. Suppose, however, that you could create a tomato, for example, that always looked and tasted just as delicious as if it had been picked out of your own summer garden! Besides, tomatoes are nothing to sneeze at financially, being valued at more than $3.5 billion annually.[47]

Tomatoes were targeted because they are the most likely to be ruined, whatever the season. If they are picked too late they get soft or rot. Picked too early and you must wait, or you're stuck with tasty but monotonous fried green tomatoes.[48] Enter Calgene, a genetic engineering company that in 1991 produced a tomato that would ripen but not get soft. By tinkering with only one of many genes, Calgene produced the Flavr Savr tomato that would ripen and retain its color and firmness without getting soggy, soft, or squishy.[49] In 1992, the U.S. Food and Drug Administration issued its first approval of a fully genetically engineered food, the Flavr Savr tomato. Genetically, the tomato-softening gene was removed and the Flavr Savr gene added.

It did not take long before every organic-loving, sandaled-footed protestor in America loosed his or her fury against Flavr Savr. Unfortunately, the protests centered very little on science and a great deal on fear. Fish with tomato heads or vice versa became common among the scare-mongering tactics. By 1994, Calgene felt as if it had been hit by every overripe tomato ever grown. The Flavr Savr tomato survived these protests but it failed in other ways. In Florida and California, the two largest tomato-growing areas in the United States, the new tomato proved to be disease-prone. Furthermore, while it might not ripen to sogginess, it could not withstand shipping. Many boxes of the tomatoes arrived bruised and soft. By 1996, the Flavr Savr tomato had lost all its savor.

However, there can be no denying the improvements side of the ledger for genetic modification if we define those improvements in terms of yield. It has been estimated that genetic modification alone is responsible for 50 percent of the harvest yields in the twentieth century.[50] For example, over the last 30 years, wheat yields have increased 115 percent; maize, 320 percent; soybeans, 112 percent; and tomatoes, 413 percent, to name only a few. But scientists are divided over the 50-percent figure. The wheat increases resulted from cross-breeding the best performers until the highest-performing hybrid was established. Some argue that this does not fall under the heading of genetic modification so much as it does under the Mendelian science of heredity. But even these techniques, which do rely on some level of genetic tinkering, are not without criticism. For example, all the cross-breeding may produce, down the line, offspring that is highly susceptible to disease, thus wiping out huge fields of wheat overnight. Others worry over the reduction in variety or the continual decrease of variety until no more can be established— an end of the line, so to speak.[51]

Yields, however, are only part of the story. Genetically modified plants are less susceptible to common diseases (but not necessarily to new ones), more resistant to heat or cold, drought, or flooding, more resistant to parasites or weeds, and very nearly generate their own inherent pesticides that are deadly to bugs but harmless to humans. On the other side of the coin, there is the possibility of mutations; we do not yet know what prolonged gene-tinkering will yield by way of mutations. Monsters? Some scientists think mutations are possible but not on a very large scale or not as species-changing as it was once suspected.[52] Other hopeful prospects in gene modification include oil recovery and microbial mining (the use of microbes in waste management and bio-leaching in mining).[53]

What has revolutionized genetic modification in recent decades has been the ability to turn genes on and off or, rather, to trigger various switches. This allows geneticists to be able to fine-tune various genes and increase desired traits or decrease less desirable ones.[54] We've known since Mendel that traits were handed down via genes. We've known since Watson-Crick how those traits were carried from one generation to the next. But research since then has enabled us to delve even further. In some ways, research into the infinitesimally small molecular structure of our very cells mimics the vastness of the starry heavens. Our molecular selves contain a vastness of information like so many twinkling stars. In the case of plants, scientists are able to change one gene and make a plant more resistant to frost; trigger another, and make it distasteful to its known predatory insect(s). But, like outer space, we are also coming to understand that the more we know, the more we come to understand that there is much we do not know.

For example, as one writer has a put it, "To an agronomist, developing plants with a genetic resistance to weed killers seemed a particularly forward-thinking thing to do."[55] Since plowing and cultivating kills earthworms that keep soil fertile and causes the topsoil to erode, what could have been wiser? But this new technology is indiscriminate and now we're killing bugs we don't want to. Furthermore, we don't honestly know what other balances we're upsetting in this process.

What's more, now that this process is more widely known it has brought out its own army of protesters. Now, protesting is the height of free speech, the bedrock of the First Amendment and a foundational tenet of what makes America, America. The problem is that protestors have become as indiscriminate as the genes scientists have created. We are now at a point of losing the baby with the bathwater. The question isn't, "Are the scientists right or are the protestors right," but, "Is there a compromise, a middle way, that allows scientists to continue their work safely while still making progress?" On the one hand, protestors say nothing is safe enough; on the other, scientists say current regulation threatens to shut down their work entirely. Neither side is without its own conflict of interest.

As mentioned before, genetic modification of plants and animals has been going on for decades. The question arises, "Why so much concern now?" If we have been eating genetically modified foods for a decade or more (and we have), why is there so much controversy about it now? And why, for example, has much of Europe come to be foursquare against it? (In Austria alone, 1.2 million citizens—about 20 percent of its total population—signed an outright ban on genetically engineered food in 1997.)[56] Indeed, the response of the European Union (EU) is a matter of separate concern and examination.

THE EU RESPONSE

The response of the European Union is linked inextricably to the growth and popularity of genetically modified (GM) foods in North America, specifically the Untied States. GM crops were initiated in 1996 and by 1998 they spread in earnest. The United States, Canada, and Argentina had grown an estimated 68.7 million acres of GM crops that mushroomed to nearly twice that by 1999.[57] Monsanto led the way in production with its Roundup Ready soybeans that contained genes for tolerance to the antibiotic ampicillin, and also with its modified maize (corn). The EU approved it. Both grains (with and without modification) made their way into Antwerp in 1996, but Greenpeace organized a wide-scale protest. It didn't work. What it did do, however (and this may have been the Greenpeace intention all along) was raise public awareness. The Germans were unhappy about what they considered soybean contamination and sought its removal, though not from all EU countries.

By 1998, Monsanto had gotten the message and launched a gigantic public relations campaign in Great Britain, touting all the health benefits of GM crops.[58] This was probably premature, as subsequent studies have shown that GM foods, like naturally grown ones, are not without their own inherent health-related problems.[59] Did this appear to EU consumers the equivalent of protesting too much? It's impossible to say, but the protests of Greenpeace and others reached a fevered pitch and soon the United Kingdom, Germany, and France had joined in the protest, angered over what they called the contamination of the food chain. Nongovernmental organizations (NGOs) also entered the debate and gained a great footing, the chain of regulation being very different in the EU than in America. In Europe, the sale of all so-called novel foods (foods that have been genetically modified) is controlled by the EC Novel Foods Regulation. While the Regulation is recent (1997), France, the Netherlands, and the UK had similar systems in place long before the EU formed. Both the EC Novel Foods Regulation and the pre-novel foods process require labeling.[60]

By the end of the 1990s it became clear that Europeans wanted to know what was in their food. They overwhelmingly favored labeling.[61] If GM foods were to have any success in the EU at all, manufacturers were going to have to label their foods. The issue of labeling will be addressed below, but what is interesting is the reaction of manufacturers of GM foods: they opposed it consistently. Even in the United States there is strong support for labeling, yet under the Clinton administration, labeling remained voluntary.[62] The blame cannot be saddled solely on GM manufacturers. Greenpeace and other protest movements do have important information to share. What dismays, however, is the proclivity of Greenpeace and its allies to rely almost exclusively on scare tactics rather than facts. For example, at nearly every Greenpeace protest regarding GM foods, someone will be wearing an outfit depicting either a fish head on a mammal or some other preposterous combination. It seemingly does not matter that no such combination has been tried or will be tried. Yet the fear that such monsters might appear is enough for the Greenpeace crowds. This is unfortunate, for it forestalls any serious debate about the safety of GM foods when misinformation overreaches facts.[63]

While most Europeans feel that biotechnology would likely raise the prospects of their lives (for example, xenotransplanation), when it comes to GM foods the overwhelming European response is negative. Spain, Portugal, Finland, and Ireland are the only EU countries where GM foods are supported at all. Indeed, as late as 2002, Europeans were still not only hesitant about GM foods, but outright opposed to them. On this side of the Atlantic, however, the perception is quite different. While some oppose it, governmental regulations, as indicated above, range from relaxed to purely voluntary.[64] The EU response will be hard to reconcile with the U.S. position, where GM foods are at best tolerated, at worst are accepted due to a lack of knowledge, stemming from the debate over the safety of recombinant DNA and its addition to seeds and food.[65] By and large, Europeans feel the Americans are moving too fast, are only concerned with the commercial possibilities, and are not forthcoming enough about when crops and other plants have been genetically modified. Yanks respond that Europe is reacting hysterically. An agreement on rules and regulatory practice would go a long way in bringing the two sides closer together.

REAL OR SYNTHETIC

According to one researcher, Glenn McGee, this is a matter of not only manner but also degree, similar to, in his words, the difference between an album, a CD, and Napster.[66] In other words, the change has been dramatic enough to alter the way we think about what it is we have before us. It's one thing, for example, to plant a rosebush that is hardier or that will effloresce more abundantly. It's quite another to sit down to eat a potato that has never been in the ground, did not touch the soil at any time during its growth, or is the product of some laboratory rather than a farmer's hand. McGee thinks it is merely a difference in technology. An album, he argues, is vinyl and Napster is software. There is no difference in the music, the end product of both. If the record (no pun intended) sales of Apple's iPod are any indication, McGee is right, at least as the analogy relates to music. Where his argument seems to fail, however (no matter how much reassurance for some) is regarding what we take into our bodies. If I'm eating the equivalent of software, it will matter, unless, perhaps, I'm on a voyage to some distant planet.

What troubles others more is the increasing control of food and food production left in the hands of a few. As the technology develops, and as we are able more and more to control nature either by accelerating, retarding, or simply changing entirely what it does, the ability of a few to control what the majority eats becomes increasingly troublesome.[67] As significant as this tension is, some people believe it is held in check by the benefits that stand in the offing. In addition to those benefits mentioned, there are others that come as a result of moving to more scientific production in farming. While increasing production and the ability to farm in places heretofore impossible, we also reduce the need for fertilizers and thus will free up oil production once needed for the manufacture of fertilizers.[68] So wouldn't the solution be to simply label everything that is genetically modified, whether it be seeds, crops, or plants that could be transplanted? The answer isn't as simple as it seems.

LABELING

Is there a moral or ethical reason for labeling? The issue emerged in 1987 with the induction of BST (Bovine Somatotropin, also written as sBST) into milk.[69] BST and chymosin rennet were introduced in the UK without fanfare, BST milk and natural milk from dairy cows not injected with bovine growth hormone (BGH, also written as rBGH; BST and BGH are often used synonymously).[70] The sale of milk came from both BST and normal sources; the sources were not separated. Neither suppliers nor buyers nor even the trial farmers were known to any other than those scientists involved in the testing. Perhaps a double-blind trial was being sought but the end result was that labeling was felt to be the fairest approach.

But labeling isn't so black and white. For example, should a cheese pizza be classified as genetically engineered if the milk used to make the cheese has been modified? To what level do we go to distinguish how a given food will be classified as genetically engineered? Is a food genetically engineered if only one part of it is, or are more than two requirements needed in order to qualify something as a genetically modified food?

One key reason for the dispute about labeling is the risk assessment made by experts (scientists) and how this is translated to laypersons (you and me).[71] What a scientist may think is low risk may not be what you or I would agree to. Most of this difference resides in the difference between a scientist's and a layperson's understanding. Even so, since it is hard to establish a level of risk acceptable to another person, some labeling might be helpful and then consumers could make up their own minds about how much risk they are willing to accept.

Some argue that labeling is so much fear-mongering.[72] Everything we eat, goes the argument, has been genetically modified. Moreover, even if you buy organic, it's unlikely that you have avoided genetically modified food because the seeds were likely "contaminated." To avoid all risk would mean eating only what you grow yourself—always—assuming you can get seeds that have not been tampered with. Any supermarket, any restaurant from fancy to fast food, any lunchroom, any hospital, or any airline offers some genetically modified food. By 2000 it was estimated that as much as 75 percent of all supermarket-processed or fresh foods would test positive for genetic modification. Labeling of any kind would have to include baby foods, baking mixes, cereals, cooking oils, corn and corn chips, corn sweeteners, dairy products, margarine, papayas, popcorn, potatoes, radicchio, salad dressings, soy products, squash, and tomatoes.[73] To escape GM would mean escaping from the planet.

Moreover, say these opponents of labeling, there is nothing to fear anyway. The U.S. Food and Drug Administration has determined that there is no information that would distinguish foods naturally grown from those grown using rDNA technology.[74] To label these foods would be pointless.

On one level, labeling may be easier by virtue of the farm conglomerates that have taken over farming. For example, in 1959 Sri Lanka grew 2,000 rice varieties. Today, only five. Eighty percent of all pork comes from only 10 percent of the pig farms, while in Greece, wheat diversity has declined by 95 percent.[75] We do not know now whether this uniformity is good. Conglomerates make identification limited to but a few; they also place most of the power in the hands of a few. For those living in Third World countries or Third World conditions, their agricultural success depends on free governments willing to provide them with their agricultural needs. In the United States, it's the same. By 1997, 30 percent of farm sales came from family-sized farms while over 60 percent came from corporate farms or proprietorships.[76]

Is genetically modified food inevitable? According to some experts, it's still possible, to some degree, to produce and use natural seeds.[77] The trouble is, of course, incentive, whether monetary or otherwise. Millions of dollars are at the disposal of anyone wishing to make a better genetically modified mouse-trap, so to say, but not much for those who want to use better parts for the traditional mousetrap. In other words, farmers (at least in the United States) find it much easier to acquire genetically modified seeds and/or crops at reduced prices than they are able to find and afford those that are not. Since 1987 and the first U.S. Department of Agriculture (USDA)-supervised genetically modified plants, acreage for GM plants has grown exponentially from 6 million acres in 1996 to 74 million acres in 2000 (109 million world-wide).[78] Moreover, universities are now combining their efforts to make agricultural–biotechnology tools far more widely available.[79] This, in turn, has led to concerns that the industry drives the university and not the other way around. While this certainly helps farmers to acquire seed-farming tech-nology, it has also been fodder to groups that oppose the genetic modification of plants.

For those who oppose genetically modified foods for whatever the reason—religious, moral, or ethical—there are always organically (or biologically) grown crops, but they have to be certified as organic foods.[80] Purely organic means that farmers have not used genetically modified seeds, a requirement that is becom-ing increasingly difficult to meet (see below). In the United States, certified organic refers to organic food that adheres to the strictest of standards. Still, a rundown of food brands testing positive for genetic modification makes avoid-ing them appear impossible: Aunt Jemima, Ball Park Franks, Betty Crocker, Boca Burgers, Bravo Tortilla Chips, Duncan Hines, Enfamil, Frito-Lay, Gar-denburger, General Mills, Green Giant, Heinz, Jiffy Corn, Kellogg's, McDon-ald's, Morning Star, Nabisco, Nestlé, Old El Paso, Ovaltine, Post, Quaker, Similac, and Ultra Slim.[81] While not every product made by these companies contains genetically modified foods, any with corn, soy, dairy, or meat products are almost certain to test positive for them.

GENETICALLY MODIFIED FOODS: THE GOOD NEWS

Genetically modified foods have nearly unlimited potential for good, or so say proponents. Take, for example, genetically engineered vitamin A rice. Vitamin A rice has been called "a breakthrough in efforts to improve the health of billions" of the less well-off, most of them in rural sections of Asia.[82] The process that creates vitamin A rice involves the insertion of a gene from a daffodil into a strain of rice to produce rice with an overabundance of beta-carotene. Beta-carotene is converted in vitamin A, which is beneficial for eyesight. The golden rice, as it's often called, is very good news for the world's poor, for whom malnutrition leads to inevitable bad eyesight or blindness or even worse.[83] Monsanto and other companies are producing canola (rapeseed), corn, and cotton that are resistant to glyphosate or the herbicide glufosinate. The idea here is to make the crops herbicide-proof so the herbicide kills only diseases, the pesticides only insects, and the crops remain healthy and growing.[84]

Other companies are making crops pest-resistant. Success has not been stellar in every case and, as we shall see below, creates its own concerns. Even so, the crops turn out to be very farmable in the most unlikely of regions. To increase the arable land in the world by even a small percentage (say, 3 percent) would mean billions would have access to more and better foods than they have now. Many proponents argue that these new breakthroughs will allow those living in the Third World to escape hunger forever.[85] Because of genetic modification, crops are now disease-, cold-, drought-, and even salt-resistant (though with very limited success), making them not only more cultivatable in soils with high salinity but also quick to mature and more readily adaptable.[86] In addition to health concerns, however, much debate continues around just how ecologically safe these crops are. Others contend these genetically modified crops are perfectly safe and the scare-mongering should cease.[87] Moreover, if crop safety is based on crops domesticated in the area of production where they are grown, then none in the United States or Canada qualifies.[88] In other words, the United States is indebted to others for having developed and domesticated its crops. Now that we have the know-how, it is our turn to use the genetically modified crops we have stabilized to help feed the world's hungry, or so say proponents of genetic modification.

Others tout the advantages that genetic modification has brought to wheat and, consequently, to bread making, to brewing using advanced barley, to yeasts, tobacco, wine, and other wheat by-products.[89] (In the case of tobacco, a gene from a firefly was inserted into the genes of a tobacco plant, resulting in tobacco leaves that glow in the dark.)[90] For example, the ability to control mildew in many cereal crops would reduce the cost to farmers by millions. The farmers' savings, the argument goes, will surely be passed on to consumers. But it isn't only agriculture that stands to gain from genetic modification. In addition to the food industries, the cleaning, textile, pulp and paper, leather tanning, oils and fats, and diagnostics and testing industries all stand to gain because various techniques either streamline or replace messy processes or procedures.[91] Some people say that even if we accept all the dangers and cautions about genetically modified crops, we must come to understand that they are here to stay. In the end, when you add up all the benefits on an imaginary ledger and compare them to all the potential dangers (real or imagined), you still end up with the benefits of genetically modified crops far exceeding any risks.[92]

Proponents of genetic modification of crops complain about the various myths circulating in the press. For example, the monarch butterfly is not only not at risk from *Bt* corn, but it has rebounded marvelously after a slight decline (although as late as 2001, some were still reporting about the nonexistent risk). Other arguments against genetic modification that surround this debate include the following: genetic modification creates superweeds; tomatoes have fish and rat genes inserted in them; normal foods do not have genes; genetically modified foods are completely unnatural; gene insertion is the equivalent to flying blind; and poor farmers in poor areas will suffer the most if genetically modified crops are completely unregulated.[93] The proponents of genetically modified foods are right about most of these myths. But, as we shall see below, both proponents and opponents carefully choose their myths to debunk.

GENETIC FOODS: THE BAD NEWS

Although there is great debate over the benefits of agricultural genetic modification, it pales in comparison to the near-incendiary exchanges about the dangers of crop genetic modification. For every claim about a danger there is a counterclaim that the danger is false, misleading, or otherwise in error. Unfortunately, the tenor and tone of those bringing the claims of danger are not always careful about facts, or are rather careless about claims.

One claim that cannot be gainsaid, however, is the claim about the infiltration of genetic modification where it is not wanted or is not supposed to be. For example, insect-resistant corn has turned up in taco shells and corn chips, exactly where it was never meant to be. Piglets genetically modified to produce less manure were killed and destined for the incinerator, yet they ended up in poultry feed consumed by chickens and turkeys, which were then eaten by humans. A farmer in Canada was fined for growing Monsanto's Roundup Ready canola, only he never planted the seeds—they blew into his fields from neighboring farms sanctioned to grow them. Meanwhile, an herbicide-resistant canola farmed in Texas cross-bred without permission with another, similarly genetically modified canola and then with a volunteer weed canola; this produced a perfectly herbicide-resistant plant that has taken over nearly an entire farm.[94]

The Green Revolution in Third World countries has persuaded farmers there to trade in their indigenous crops for fewer high-yielding varieties that require large amounts of herbicides and pesticides. Additionally, the Green Revolution made it culturally unpopular to eat brown rice instead of highly polished white rice.[95] Many consider this a step backwards rather than one giant step forward. If these techniques and crops do not flourish, it's argued that Third World farmers will be worse off than ever before because they will not feed the world, much less themselves.[96]

Other concerns rise immediately to the surface. Allergic reactions to genetic modification have caused everything from respiratory distress to diarrhea, rapid heartbeats, and even death.[97] One of the more serious dangers is the fear that by "transferring genes between very different organisms, scientists will create varieties with radically new capabilities."[98] The problem goes back to the gene insertion unpredictability we discussed in Chapters 2 and 3. For example, when grasses that were once burned at the end of the season were exchanged for genetically modified grasses that could be fermented and recycled, wheat grown in this recycled soil died in about seven days. Others fear what will happen to genetically modified trees when they cross-pollinate. Can they be counted on for the right mix of oxygen/carbon dioxide exchange the planet relies on? Some are doubtful.[99] Others fear that technology has either outwitted science, outpaced it, or both, while still others believe the problem has reached critical mass and are calling for a moratorium on all GM foods and crops.[100] All wonder about the number of transgenic crops we have already created and whether it's too much: wheat, rice, canola, melons, apples, coffee, squash, cucumbers, eggplants, strawberries, onions, peas, cranberries, pineapples, plums, tomatoes, walnuts, raspberries, tobacco, potatoes, sweet potatoes, and watermelons, to name a few.[101]

Transgenic crops cross-pollinate to produce wild ancestors and so create another serious problem. In every case where there is a new species, there soon follows a so-called wild relative or a novel organism.[102] Although not an exact analogy (but close), anyone from the South knows about kudzu, a vine that can grow as much as a foot a day. Some scientists are fearful that the cross-pollination with something like kudzu (which cannot be forestalled) will result in a superweed that cannot be eliminated, regardless of the tactic used. The fear is that rather than creating a boon to agriculture and the plant world, we have created an unstoppable boondoggle from which there is no escape. Even efforts to prevent superweeds via the USDA's Technology Protection System, alias the Terminator, did not work as planned.[103] Technically, the Terminator would render a seed sterile so any future seeds would, in essence, be firing blanks. Whether the name or the technology ruined this effort, all it ended up doing was giving genetic modification a black eye and leaving its creators seething. Finally, there is no agreement on just how far seeds cross-pollinate. If GM seeds are planted in one place, can they be contained? Some say no and cite as evidence GM seeds growing in places as far as 400 miles from where they originated.[104]

CONCLUSIONS

In the end, this debate is about science and how far it can be trusted. When science is the judge and the jury when it comes to genetically modified foods, public safety is in the dock. Clearly, money (and huge sums of it) drives the development of GM and its acceptance. This, by itself, would not be cause for alarm, for many things in a democratic, capitalist society are driven in just this manner to no ill effect. It becomes a problem when money is the only reason for developing or driving products and when only well-heeled stakeholders hold all the power.

We know that transgenic foods do have value. We also know that there are some inherent risks. The question becomes, how much of each? At this early date, GM foods and crops appear to be no more dangerous than normal foods grown with pesticides, or even than organic foods. We also know, however, that their dangers are serious enough to require some intervention by some agency not driven by biotechnology and not having a claim therein. The following recommendations seem to be the best course of action.

Slow down the rate at which transgenic foods are being placed on the market until more research can be done. This is not the same as a moratorium. By slowing down the rate at which these foods go on the market, we could at least be certain of how many risks there are, how serious they are, and whether we should proceed. This delay would also allow for more longitudinal studies to be done to ensure health and safety.

Do not mandate what farmers grow. We should not demand that either American or Third World farmers grow GM foods. If both or either are willing to take risks, let them, but give them a bona fide choice.

Create stricter standards for GM foods and crops. The FDA (or some other government agency) should make certain that GM products be kept out of foods where they have not been approved. Periodic testing should be mandatory and, when violations are found, fines should be significant enough to deter future wrongdoers.

Finally, *all foods should be labeled. Caveat emptor.* Let the buyer beware. If GM foods are to be in abundance in grocery stores, then consumers should know this. We already flag foods for aspartame, splendida, and even alar. Why not for genetic modification? If a consumer wants to buy them, fine; if not, then the market will have spoken.

These few recommendations will allow for some choice by consumers in the market and choices for farmers in the fields, while still presenting entrepreneurial opportunities for stakeholders. Meanwhile, both sides must endeavor to carry on this debate without malice and hyperbole. If we cannot have a reasonable debate, we will be doomed to unreasonable outcomes.

5

Well, Hello, Dolly: Animal Applications of Genetic Engineering

When looking back at the genetic engineering events of 1996, one would never guess that an innocent-looking Finn-Dorset ewe—a lamb, really—would change the face of the genetic engineering debate, but so it did.[1] The world said "Hello, Dolly" with great trumpeting (but without Satchmo) to another female, though not a woman. Dolly was not just any ewe, but a ewe "cloned ... from a cell that had been taken from the mammary gland (i.e., the udder) of an old ewe and then grown in culture." The ewe had long been dead but biologist Ian Wilmut and his colleague, Keith Campbell, "reconstructed" (their word) an embryo and placed it in the womb of a surrogate sheep. The two became the best-known scientists in the world, if only for a short time. Both sensed the coming debate and hired a public relations firm to help manage the dissemination of information.[2]

Dolly was not the first mammal to be cloned, but was the first to be cloned from an adult body cell.[3] The two scientists did not make public the existence of Dolly (or lamb 6LL3, as she was first known) until February, 1997, and they did so in the science magazine, *Nature,* known for its groundbreaking scientific discoveries. Until that moment, no scientist thought the process possible or, rather, perhaps thought it possible but highly improbable. Dolly ushered in a new age and a new and highly controversial twist to this already disputatious subject.

TRACY, MEGAN, MORAG, DOLLY, AND POLLY

Wilmut did not stumble upon Dolly accidentally. He had already worked on Tracy, not a fully cloned sheep but the first transgenic animal with commercial significance. Tracy preceded Megan, who came before Morag, who led to Dolly.[4] The point? Only that Tracy, who was born in 1990, came long before Dolly (1996) and represented a great deal of work that led up to Dolly, who stunned everyone but only because they had not been paying attention. Not even Polly, Dolly's successor, had anywhere near the impact or press that Dolly had and continues to receive. The advantages of being the first of anything tend to crowd out the rest as ho-hum, even for something as terribly unho-hum as cloning!

But why was everyone so surprised at the so-called impossibility? One reason why it was thought to be impossible was because of gene imprinting.[5] In common language, this means that it really matters from which parent (your mother or father) you get certain genes. In the case of certain diseases, for example, getting them from your father (or your mother) will mean the disease will be remarkably severe or mild. The process that produced Dolly (nuclear transfer) made the gene imprinting phenomenon known (through Azim Surani's work). Until Dolly, giving mammal zygotes two female or two male pronuclei meant they would eventually fail when implanted in the uterus.

Dolly helped scientists see their way around this problem and create a cloned animal through nuclear transfer, even one whose offspring (in Dolly's case, Bonnie) still lives. Until the advent of the nuclear transfer cloning technique, most saw gene imprinting as an impossible obstacle to overcome. To get around it, however, another technique that would lead to even greater controversy had to be invented. More on that later.

In this line of remarkable sheep, Tracy provided something especially important. She had been fitted with a human gene that produces the enzyme AAT (alpha-1-antitrypsin), making her milk-producing abilities legendary. But AAT is more important than that; it is the same enzyme used to treat emphysema and cystic fibrosis and other lung diseases.[6] Until Tracy, the only way to get this enzyme was to extract it from human blood plasma via an expensive process. Since anything human also has human infections and diseases, some scientists thought another avenue was needed; thus, Tracy et al. Clinical trials of AAT gene replacement therapy should begin within the decade.[7]

This treatment led to one primary reason, at least according to Wilmut and his colleagues, why Dolly came to be in the first place: cloning holds many medical promises. Although most such promises have been disappointing, hope springs eternal. Wilmut himself has argued that cloning technology will help treat diseases by producing proteins, by producing organs from animals that can be later transplanted into humans, and by producing human cells than can be inserted to replace damaged ones.[8] If you or someone you love struggles with a disease that might be cured, or at least significantly assuaged by genetic engineering, you most probably want to see it continued. If that were the only matter to consider, there most likely would not be any debate, or certainly not like the one we now have on our hands. As the old saying has it, however, it takes two to tango, and there is another tango partner in this debate.

The controversy involves the requirement to culture inner cell mass (ICM) cells in order to help them retain their totipotency.[9] This means taking them at a young age, before they have been switched on or off to become a certain tissue. In other words, it means taking them at the embryonic level or—as you might have guessed—before they are born. Wilmut believed (and believes) that this can only occur at the embryonic level, though a scientist named Jim McWhir achieved it with frogs at the postembryonic stage.[10] But frogs are not mammals, hence Wilmut's efforts to achieve totipotency in mammals. Few have concerns about this for animals (except those who protest for People for the Ethical Treatment of Animals, PETA). But start talking embryonic stem cell lines (ESs) from all mammals (i.e. humans) and the debate rages quickly with many more opponents.[11]

Wilmut and his colleagues began this process with two earlier sheep, Megan and Morag, mentioned above. What they discovered is that cells could be reprogrammed and taught to regain their totipotency.[12] Megan and Morag showed Wilmut and his colleagues as early as 1993 that this was possible by proving that the cell cycle is much more important than cell differentiation (perhaps the former is the most important aspect of all). In other words, Megan and Morag showed the world (though the story was carried mainly in the United Kingdom) that the cloning of mammals is possible from adult cells if those cells are forced to enter a resting phase only to exit the growth cycle and then enter it before the gene transfer.[13] The reception of Megan and Morag outside scientific circles mimicked what was said in Chapter 1, mainly that what was being done here could either produce great miracles or great monsters (as one headline actually ran). For the scientists, Megan and Morag proved more interesting scientifically than Dolly, but Dolly proved more PR-worthy.

With the technique of cloning a mammal now in place, Wilmut and his colleagues began in earnest on Dolly. What made Dolly so headline-grabbing is that she came from an adult cell. But Dolly did not come into being immediately. Wilmut and Campbell took 277 embryos from the Finn-Dorset mammary cells but recovered only 247. Of these, only 29 began proper development and they were implanted into 13 ewes. From the 29 implantations, only one ewe became pregnant and she gave birth to Dolly. Certainly the odds were against them—1 out of 277—and one can only reflect what would have happened if they had tried 1 out of 276 (with the successful Dolly left out). On the other hand, it really is only 1 out of 13 and this rate is about the same when in vitro fertilization is used in humans.[14] Still, the rate of success was then, and continues today, to be a bone of contention because it requires so much to gain so little. The lamb that followed, Polly, proved that all the biotechnology necessary for Megan, Morag, and Dolly could now be relied upon as scientific fact—that is, Dolly was no fluke. Polly came just in time, too, as the British government ended its grant to Wilmut, forcing him to scramble for private funding.[15]

So why are all these sheep so important? The obvious need not be mentioned but here it is: this technique, considered impossible for 100 years, is now possible. It reminds one of the circulation of blood. The Greek scientist Galen (c. 200 B.C.), the first who gave it serious thought, believed blood was made new with each heartbeat. Centuries later, William Harvey (1578 – 1657) determined just how the blood flows (i.e., that it was recirculated and not made new each time the heart beat). In the case of Dolly, cloning has now become a fact of scientific biotechnology. Only three or four decades ago, heart surgery was thought to be impossible. Today it is routine. With Dolly now on the scene, many are thinking that cloning will eventually become routine, and not a few are bothered by this.

Wilmut and his colleagues believe that cloning is important for medical breakthroughs.[16] Imagine having two or three sheep that different drugs could be used on to determine the best one. This is the idea that Wilmut and his colleagues had. But what bothers some is that Wilmut is not the only scientist involved in cloning, and not all of them have the same altruistic motives (though Wilmut patented his cloning techniques, of course). Below we will examine what some of the problems are and why more than a few scientists and laypersons are concerned about cloning and about its eventual acceptance.

What remains on everyone's mind is, of course, human cloning. Wilmut and his colleagues "see human cloning as a rather ugly diversion, superfluous as a medical procedure and repugnant in general."[17] But they also have made the idea much more scientifically likely by showing how it can be done with mammals. Wilmut and others are quick to point out that while they find human cloning repugnant, they cannot tell the rest of the world what to do with cloning or what to clone.

Indeed, recall from Chapter 3 one Richard Seed who promised to clone a human by 2000.[18] We do not have one yet and that brings up the question of how long it will take. Doubtless a good while, but so did artificial insemination, once thought impossible since male ejaculate had to be fresh. But techniques in artificial insemination allowed for the freezing of ejaculate, making it a routine procedure today. To argue that human cloning cannot be done is to miss the story history teaches. To argue that it should not be done is a more reasonable, ethical debate. If we move too far in that direction, the movie *Multiplicity*, though technically silly, may well become a future reality that does not look too appealing.[19] Indeed, Huxley's hatcheries look more and more like a prognosis rather than a fantasy.

Human cloning in the United States has been all but stopped by Congressional legislation. HR 923 forbids federal funds being used to clone humans and also levies a small fine ($5,000). More than a dozen states have or are going to pass similar legislation and have called upon President George W. Bush to make it a federal offense.[20] Some have raised serious ethical concerns regarding whether cloning jeopardizes the moral and human right to a unique identity.[21] Others have weighed in on the theological side of the question, asking first whether we have this right, and second, whether we somehow short-circuit the soul in this process. Most prominent in this debate is Leon Kass, President Bush's point man for his administration on cloning and genetic engineering. Kass led the ethical side of this discussion long before he was appointed head of the President's Council on Bioethics.[22] He has led the discussion with intelligence and vigor. Kass has been pivotal in quashing human cloning and now most experts agree that cloning-to-produce-children (CPC) is too unsafe and should be prohibited.[23] With the exception of people like Seed, human cloning will likely gain little traction but the debate will rage on.

But the cloning bans did not get underway until hoax after hoax plagued the science of cloning. Odd in the debate has been a book by David Rorvik, *In His Image: The Cloning of Man* (Lippincott, 1978). He claimed that he had participated in the cloning of a reclusive millionaire.[24] The story turned out to be a hoax and gave science a black eye, but not before it made bigger headlines than Dolly. Although this book came out long before mammals had ever been cloned, it received a credible reception even though its author could claim no other credentials beyond those of a freelance writer. The Rorvik episode underscores the necessity of examining every claim thoroughly.

Six years after Dolly, a weird, quasi-religious cult known as the Raelians (they believe they have been tapped to welcome back extraplanetary visitors) claimed to have created the first human clone, a young girl named, appropriately enough, Eve. A willing press and a mostly gullible public ("Eve"?) held onto the story until major media figures such as Connie Chung had disgraced themselves.[25] Rael is the former French journalist Claude Vorilhon and leader of the Raelians. He claims to be a direct descendant of extraterrestrials who created life on Earth using genetic engineering. Bridgette Boisselier is the scientific director and chief executive officer of the Raelian-founded Clonaid. When both Rael and Boisselier refused to allow independent examination of Eve or her mother, the ruse became fully known.[26] A little science would have put this story to rest almost immediately. When the story became public, no human clone had ever gotten past a few cell divisions.

Suppose, however, we set aside human cloning and look only at animal cloning. Does that change the debate? Some, but not much. While some have argued that animal cloning would lead to more and better animal-to-human organ transplants, others, such as PETA advocates, argue that creating animals for the sake of killing them violates every known ethical principle. Yet animal-to-human organ transplants could save thousands of people who have heart, kidney, or stomach diseases and even offer hope to AIDS patients.[27] Others point out that transgenic animals can function as living test tubes or provide a means for preventing the extinction of species.[28] While all of this may be true, very little of it touches upon animals and their pain.

OTHER TECHNIQUES

To date, cows, sheep, goats, pigs, insects, mice, birds, chickens, deer, and fish, to name a few, have been cloned or created transgenetically. All have provided grist for the cloning mill.[29] A number of techniques have become commonplace, in addition to nuclear transfer. Nuclear transfer uses small needles and micromanipulation to insert a nucleus and DNA into an enucleated egg.[30] Most scientists see this as the preferred technique for cloning mammals for now. Other animal cloning techniques (broadly defined), while not always yielding a clone, do figure into the process scientifically, if only peripherally. Transfection allows for the insertion of DNA into animal cells, first into the cells (actual transfection) and later into the genome.[31] Chemical transfection allows for the insertion of DNA in the presence of cells.[32] Embryo splitting involves taking a young embryo and dividing it into halves, thirds, or quarters and encouraging the divided embryo into development. It yields twins, triplets, or quadruplets and has been going on much longer than nuclear transfer.[33] Fetal cell cloning creates a clone from totipotent cells, while stem cell cloning utilizes stem cells that develop very early in the life of the cell. In every case, the number of clones can be very small (in the case of Dolly, only one; in the case of other techniques, three or four).[34]

Electroporation subjects cells to an electric field, causing them to process new DNA faster.[35] Still other techniques such as the Capecchi technique, while not yet perfected for the cloning of animals, are being worked on to make them so.[36] While Dolly remains something of a scientific miracle, none of the techniques has high yield rates and they continue to contribute to ethical dilemmas. Given their rate of return, if any of these techniques do in fact work on humans, the ethical debate would turn from warm to incandescent. Would we want to sacrifice more than 200 human clones to get one successful replica?

Although transgenic animals are not the same as clones, they can be discussed in the same breath, as it were, because transgenic animals are animals with extra DNA that is not their own, or did not come with the DNA nature provided. Scientists often use zygote injection to accomplish this. This, the most efficient method, allows for the injection of DNA into the zygote and then allowing that zygote to develop in a pseudopregnant mother (an asexually pregnant mother).[37] This technique is important because it allows scientists to examine cells and cell growth while allowing them to isolate various cells and DNA for observation in an isolated context. It requires that the zygote be removed from, say, a mouse (the most commonly used animal) that has mated the night before and, using a microscope, it is inserted in the male pronucleus. Later, the zygotes are transferred to the oviducts of another female mouse. This process usually yields a 50 percent survival rate, with only 20 percent making it to term. Another procedure allows for the literal mixing of cells from two different embryos, producing what it called a chimera.[38] This allows for various kinds of study that would not otherwise be possible. While the success rate is higher, many of these animals do not survive for long and most experts agree they do suffer pain, though there is disagreement over how much.

Animal testing has, of course, been done for decades but the advent of Dolly allows for more extensive testing, the likes of which we have not had access to. The reasons for this were alluded to earlier; they allow for scientists to do all sorts of tests in what are identical specimens to see what happens for the benefit of either other animals or humans. In addition to the diseases mentioned above, cancer treatments are likely to benefit, along with the tragic Alzheimer's disease, atherosclerosis, type I diabetes, hypertension, and what is called apoptosis (cell death). Cell death leads to tumors, the decline of the sensory system, and autoimmune diseases. By marking and working with genes in the cloning context, or separately, scientists have made tremendous headway in the pursuit of treatment or cure of diseases.[39] Again, however, how much headway is debatable. No one doubts we are farther along than we were only ten years ago, yet there remains no real change in the cure rate of any of these diseases compared to any time before these techniques were implemented.

Some view this as animal agriculture and favor it because of the potential or already realized benefits to mankind. Others see it as a violation of ethical and moral behavior. People holding the latter view are often astounded by the waste of animal substance to produce a transgenic animal. For example, in one case of transgenic cows, 2,470 cow eggs were used, of which 2,297 matured, 1,358 were fertilized, 1,154 were injected with human DNA, 981 survived that procedure, 687 began embryonic development,

129 were transferred into cow oviducts, 21 cows became pregnant, 19 calves were born, yielding only 2 actual transgenic bovines.[40]

Even the U.S. Patent and Trademark Office issued a self-imposed moratorium (later repealed) on three kinds of transgenic mice because of the potential pain they surely suffered. For example, the genes of the very famous Oncomouse have been genetically modified so it is highly prone to cancer. That process was later patented.[41] The question becomes one of whether we should sacrifice animals to save humans from common diseases, or give animals equal status with humans and let them both die as nature requires. While the answer will seem obvious to some, others are not so willing to agree. In the abstract, this question is easy to answer. Imagine, however, having a child whose terminal disease could have been cured, or greatly abated, if only a certain gene therapy could have been introduced.

Another transgenic mouse, referred to as the Human Mouse, has been around much longer than Dolly (it was created in 1988) but is still making news, though again not as big a splash as Dolly. The Human Mouse has the immune system of a human via gene insertion.[42] The mouse allows for various kinds of experiments involving drug interactions with the human immune system. Obviously, such procedures would be ethically impossible on humans today, so these mice provide great research venues.

ANIMAL CLONING: PROBLEMS AND PROSPECTS

Research into animal cloning and transgenesis continues. Techniques to contain the overexpression of milk in mice may have some future application in farm ruminants, such as reducing the potential for lactose intolerance. Increasing the water in milk (by reducing the lactose) would also reduce mammary engorgement in cows while reducing infections. By overexpressing certain growth hormones in transgenic pigs, scientists have been able to increase overall body size. Using the same technique in fish (e.g., salmon, carp, and catfish) has also increased their size, as well. In the case of salmon, for example, the result is better, fatter, meatier salmon.[43] Using a similar method in transgenic sheep has increased their wool growth.[44]

Although the techniques are certain to improve, reproductive cloning is not foreseeable for the reasons stated above: low yield and high cost. What will doubtless occur, however, is the reproductive cloning of unique or endangered animals. Even though the yield is low, gaining time for these species (even two or three at a time) would be advantageous in the minds of many. At this writing, research into gene therapy from animal research remains inconclusive. While infertile mice have been treated by gene insertion to become fertile, targeting specific diseases remains, at best, a shot in the dark.[45] Foreign gene insertion in animals still results in serious side effects and even death. Obviously, more research is needed, though some would argue that given these deleterious side effects all transgenesis should cease immediately. The insertion of dangerous or potentially deadly pathogens into transgenic animals still raises many concerns, all of them harking back to the Asilomar debates covered in Chapter 3. Can these procedures be followed without any possibility of more widespread, unintended infection and consequences, or are they too dangerous to attempt? Most, though not all, scientists say yes; others are less optimistic, less certain, and they lean toward a moratorium.

Bovine growth hormone (BGH, also referred to as BST, or bovine somatropin), mentioned in the previous chapter, requires an additional comment here. Recombinant BGH (rBGH) is injected into cows and mocks a natural growth hormone that increases milk production in those cows.[46] It is not a small matter of increase, either, with estimates as low as 10 percent and as high as 20 percent or slightly more. At first blush this may seem like a small number, but put it in context: if you could increase your income by 10 or 20 percent, would you be interested? That's how some dairy farmers feel.

The debate about this process arose when it was discovered that BGH had to be spliced into our old friend, E. coli, the well-known bacterium. Before all the data were in, and even nearly a decade before the FDA ruled that BGH was safe for cows, the FDA agreed that BGH was safe for humans. The FDA argued there was no difference between natural milk and BGH-enhanced milk. But BGH-enhanced milk does have a molecular structure different from natural milk and it does cause some problems in cows that are mimicked, or can be, in humans.[47] Before the matter ended, BGH had become the protest du jour. Given the stridency of the debate and its extremes, BGH-enhanced milk could be as harmless as water or as deadly as bubonic plague.

The latter proved to be so much nonsense but, as we mentioned in the last chapter, labeling would have solved much of the problem. Those who do not fear BGH-enhanced milk should be free to drink it. While it does hold the potential for certain allergic reactions (not unlike natural milk does for many), evidence to date does not appear to be consequential. Some have also raised the question of why scientists should bother, since milk consumption has declined over the last two decades, indicating a possible oversupply. Others point to Third World areas where more milk is desperately needed.

What still gives many pause, however, is its long-term use. For example, we did not know that excessive amounts of fat significantly contributed to serious heart disease until millions of people had died from it. Some fear we will discover, 20 years hence, that BGH contributes to a similar serious fate for those who consume it now.

Although Canada has fought against BGH, the United States has generally accepted it. One notable exception stands out—Ben & Jerry's Ice Cream.[48] The company, known for what some have called its radical politics, has opposed BGH for some time. The opposition included a legal battle as well. Ben and Jerry have since sold their company, to their like-minded employees' dismay.

Hog ranchers have, or may eventually benefit from, transgenic or cloned enviropigs.[49] If you've ever been stuck behind a truck on the highway hauling pigs to market, you don't need me to tell you what needs to be corrected. But the stench is only part of the problem. Pig excrement, if released into waterways, causes widespread disease and fish kills. One reason for this is the phosphorus in the diet of pigs from plants and other foods. If the phosphorus or the pigs could be changed to make it (or them) enviro-friendly, then the problem would be potentially solved. Many think transgenic pigs that create their own phytase would do the trick by eliminating the stench. This has not yet been perfected but work continues. The claim of enviropig proponents is that if perfected these pigs would produce either less phosphorous-laden manure, less manure stench, or both.

Much has been made about Dolly but little about Gene, the first human-cloned cow. The technique used with Gene differed from the one used in Dolly et al. in that the cells used were far more totipotent. But this approach appeals to ranchers and farmers because the offspring tend to yield uniform meat quality and high milk production.[50] Some animal laboratories have turned their animals into so-called bioreactors, or what has been referred to as the living equivalent of such. These animals produce various desired proteins that would not naturally occur, for further research. [51] These have been referred to in the (often adversarial) press as animal pharms or pharm animals, drawing on the obvious reference to pharmacies.[52] Most recently in the animal-cloning news (summer 2005) is the first successful cloned dog, "Snuppy." An earlier cloned dog died, but after more than 16 weeks, the Afghan Snuppy is doing well. According to Woo Suk Hwang, head of the Korean research team, dogs have very similar characteristics to humans. Some canine diseases are very nearly identical to human ones as well, leading the Korean researchers and others in the science community to herald this cloning success with near Dolly-like fanfare.[53]

CONCLUSION

How are we to balance the need for research into various human diseases using animals, and the rights of animals themselves?[54] Clearly, animal research had provided groundbreaking research into particularly difficult diseases such as cystic fibrosis and emphysema, though both are a long way from being cured. How, in a pluralistic society, are we to be protective of the ethical considerations of others when they differ either from our own or even from the majority?

Answers are not easily forthcoming. On the one hand, researchers are eager to undertake this research and thereby may have a conflict of interest that prevents them from being the best judges. Yet those who are opposed to most or all animal research likewise have a conflict of interest, though it differs from those of the researchers. Setting up review boards seems like a wise alternative (though similar boards have often been ineffective or have allowed for what some consider too much latitude).

Perhaps we can focus on what we know. When it comes to treating preventable human or animal diseases, it may be best to give researchers more latitude but also employ very strict guidelines. Creating animal clones for certain death when not in the service of clear human or animal betterment should be prevented. Animal patents, or patents on life of any kind, should be strongly regulated. Like most human endeavors, when large sums of money are to be made by small groups, the risk for potential wrongdoing rises exponentially. Some regulation of possible paybacks (for example, one genetically modified chrysanthemum was named Moneymaker by its designers perhaps anticipating the hoped-for outcome) could reduce the chances of research for that end alone.[55]

Cloning meatier animals or farm animals with higher milk production is less clear ethically. The need for such animals is less obvious. Still, needs such as those for rice and other crops in underdeveloped countries raise issues about hunger and its elimination that drive this research. More study and research is needed to determine if these animals will, in fact, solve or at least attenuate the problem of world hunger.

Neither of these concerns touches on the ecological problems raised by transgenic animals and animal clones. In the same way that transgenic crops create special ecological concerns, transgenic animals and animal clones do, too. What will happen if these animals are released into the general animal population? We do not yet know and, for that reason, need to exercise great caution. Some readers may see this as easy to control in the case of animals like Dolly and Polly. But already lambs, and particularly fish and birds, have raised more serious issues about control. Accidental release of transgenic or cloned animals is still a major cause for concern until more research has been done. We are rightly fearful of so-called superweeds. Thinking similarly about transgenic animals is not irrational.

There can be no doubt that we have ushered in a brave new world with transgenic and cloning techniques. We have created special, ethical concerns with the creation of genetically modified crops and animals. Because neither crops nor animal behaviors can be said to be fully predictable, we do not know what potential dangers we have also unleashed. When weighed carefully, the potential benefits (especially in the case of human and animal disease control and cure) still outweigh the potential risks. Yet in our desire to make the world a healthier and disease-free place, we cannot close our eyes to the dangers those risks might entail.

6

Where No Man (or Woman) Has Gone Before: The Human Genome Project

Ever since Mendel first linked traits—particulates, as he called them—with heredity, man has known about the power of genes. Indeed, even before that, as we pointed out in Chapter 1, the observant noticed that by breeding animals with certain characteristics, usually resulted in getting more of one desired trait than another. But until Watson and Crick's famous discovery, most of this grouping was guesswork only slightly more accurate than Nostradamus's. Yes, it could be predicted in broad, general outlines, or no it could not be predicted with very much specificity, or even from generation to generation. Even after Watson and Crick's better mousetrap view of molecular biology, genes explained much; but until we could match genes with traits (or even with diseases) the best we'd get would be educated hunches. With the discovery of genes, the task seemed plain: find out which ones do what and you'll have much of the riddle of human life and human diseases solved.

When Galileo looked at the heavens, he declared that they were indeed the language of God's creation. It could be said, without too much of an intellectual stretch, that the Human Genome Project, with all its DNA descriptions, is God's alphabet of the creation of life.[1] When the Human Genome Project had its day on June 26, 2000, three scientists stood as its Titans.[2] Two scientists, James Watson—yes, that Watson from Chapter 2— and Francis S. Collins, and one entrepreneur and renegade scientist, J. Craig Venter, all stood with President Clinton in the White House East Room and celebrated the completion of the arduous struggle to name that gene, as it were. Collins had made a name for himself 10 years earlier with the discovery of the gene that causes cystic fibrosis. Venter, the former government researcher and now the Bill Gates of biology, had unfurled Celera Genomics, a private firm that he hoped would get the gene mapped before the government did.

All three stood there smiling, but their smiles hid what most in the room knew: the animosity among the three men was so thick and palpable it could be cut with a knife, even a dull butter knife. Venter and Watson had fought with each other for years, while Collins and Venter had crossed swords more than once. Their different views about the best way to approach gene hunting, their different ideas about each other's work, and their egos the size of small countries make this Kodak moment more like a portrait of the Gotti family. Had they posed as they actually felt about each other, it might well have looked like a scene out of *Braveheart* with Watson and Collins on one side, Venter on the other.[3]

Celera Genomics had its work cut out for it for on the day the federal government announced its investment of $3 billion in the Human Genome Project. Vying for the government's essentially unlimited repository of money (Congress can, after all, print its own cash flow, and has ever since there's been a deficit) would be a formidable task, regardless of whether your name is Rockefeller, Gates, or Celera.

The $3 billion represented about $200 million annually for the next 15 years, or the gross national product of several underdeveloped nations.[4] But Celera was there to prove that one of the main goals of the Human Genome Project (see below)—the transfer of technologies to private industry—would be carried out. The amount of money allocated to the project was immense and showed just how much importance the national government placed on the Human Genome Project. Congress had already invested in gene research long before then. Ever since World War II, the Department of Energy had a long-standing interest in DNA and its mutations.[5]

Watson initially headed up the HGP and had delayed its beginning until 1990. But he soon became embroiled in outside interests. His disagreements with other NIH scientists (such as Craig Venter, who left NIH in 1992 to launch Celera) came to a head and he finally resigned after many fiery disagreements with NIH director Bernadine Healy.[6] Venter had been dubbed Biology's Bad Boy and he continues to wear that crown today.[7] While he left the Human Genome Project early on, he returned in 2004 to acquire commercially available pieces of DNA to make simple genomes to make new life forms, a venture for which he has been roundly criticized. Collins was the logical next choice to head up the project, but he didn't want it until Healy made him an offer he couldn't refuse. The celebrated $3 billion budget Collins later put forward caused debate and disagreement inside and out the NIH, as we'll see later in the chapter.

The goals of the Human Genome Project (HGP) were simple and straightforward:

- Map and sequence the human genome, or as much of it as can be done
- Prepare a model map of a mouse genome
- Create data links between scientists
- Study the ethical, legal, and social implications of same
- Train researchers
- Develop technologies
- Transfer these technologies to industry and medicine.[8]

Some would later argue that this much money should never have been allotted to such over-reaching goals. Others would argue about the allotted amounts, but for different reasons. Note, however, the fourth goal: study the ethical, legal, and social implications of same. Clearly, geneticists had learned something from Asilomar. In fact, they had learned so much so fast that they were willing to award the project's Ethical, Legal and Social Issues (ELSI) committee 5 percent of the overall budget.[9] This award led many to argue that the direction of philosophy—toward bioethics—while moving in that direction anyway, would forever hinder its movement in any other direction. Even the $3 billion may not be enough. Current estimates indicate that base pairs of DNA are costing about $5 to $10 each. At this rate, it may cost $30 billion to complete fully.[10]

With the work underway, most scientists thought that they would have to map 100,000 human genes, which they considered a reasonable estimate. This figure came about through deduction. We know that the human genome has 3 billion base pairs of DNA. We do not know the number of genes required to make humans, but for mammals it's about 30,000 bases long. This average seemed logical. But as we pointed out in Chapter 3, this estimate was later revised to about 75,000, later to 60,000, then to 30,000 to 40,000 in 2003, to today's estimate of 20,000 to 25,000.[11] This has proven most surprising, as early estimates were based on what we already knew. Although we know humans are complex creatures, the number came as more than a shock: we are—gene-wise speaking—less complex than rice or corn (both of which have twice as many genes as humans) and only a little more complex than the fruit fly or *Caenorhabditis elegans*, a researcher's favorite worm.

Complexity (or lack thereof) aside, a map of the human genome unfolded enormous detail about humans: how we grow and develop, what makes us specifically human (as opposed to another species), and how we differ from one another and across species.[12] The second of these questions proved most enlightening, as the genetic difference between us and, say, chimpanzees, is less than 1 percent! Of course, that 1 percent sounds small until you do the math.

The difference in base pairs between two siblings is 2 million; between a man and an unrelated woman, 6 million; between a man and a chimp, 50 million; and between man and a plant (spinach, for example), about 2 billion.[13] While some researchers marvel at the smallness of the differences, others are astounded by the difference that tiny 1 percent makes. Mention has been made of the diagnostic tool this mapping would provide, since most scientists agree that every disease (with the possible exception of certain traumas) has a genetic origin or cause. But that accounts for only about 2 percent of all genes. What the other 98 percent are responsible for would also prove most enlightening, or so researchers thought. Scientists got this information using both positional cloning (locating a gene on a map and then searching through all the complex DNA until they find a difference) and functional cloning (where physiological and biochemical studies are done to figure out what is functionally wrong and then locating the culprit or gene responsible).[14] Obviously, positional cloning has become more fruitful with the advent of the genome map. For most diseases, functional cloning does not work because we know so very little about the disease itself, or even how to begin the process.

THE MAP TO HEALTH OR HAZARD?

The Human Genome Project kicked off to much fanfare and has, by and large, been received in the same manner, not unlike the celebrated Dolly of the previous chapter. Collins himself called it "the most important and the most significant project that humankind has ever mounted."[15] Others said it ranks in achievement with the works of Shakespeare, the paintings of Rembrandt, or the music of Wagner.[16] It's unlikely that the adulator was thinking of Wagner's *Ride of the Valkyries* but, even so, the road to Genetic Neverland has not been an exclusively yellow-bricked one.

Like most new technology, the Human Genome Project has raised grave ethical concerns. For example, will the poor have access to this map or will they merely be used as tools to help create it?[17] Others worry about the sequencing project and what it opens up to insurance companies or other agencies that might want to have access to this information.[18] Others have referred to it as the Manhattan Project for genetics, having both good and bad outcomes.[19] Not a few fear that such a massive federal investment has siphoned off critical dollars from projects that could be completed much more cheaply, save lives, but have a less gaudy scientific outcome.

Some fear that the federal money will be followed by private foundation money, making it nearly impossible to find scientists who can perform purely independent research.[20] Some, like Leon Kass, dispute the reasoning behind the project; he claims that it will not lead to a better life, simply to

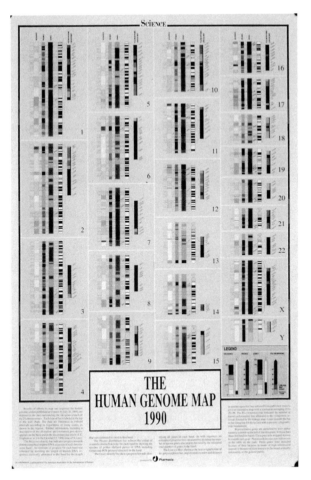

Road to Utopia, or road to nowhere? Human Genome Map.
From the National Library of Medicine, History of Medicine
Collection.

a more complex one, requiring more difficult decision making based on
that knowledge.[21] Others see the HGP as ushering in a new age of societal
eugenics and "experimentation gone horribly awry," or they refer to the
HGP as involving "big money, big consequences, and big controversies."[22]

HGP has enough good and bad involved to give everyone something to
complain about. Some have attacked one of the major goals of HGP as being
destined for failure: the transfer of technologies to private industry. Some
observers, themselves scientists, fear this perceived competition will possibly
compromise standards or place lesser but more commercially viable projects
ahead of more important but less commercially significant ones.[23]

Although these concerns focus on a number of different issues, they can be subsumed in one question: will all this money divert attention from more serious concerns? One set of concerns is macro in nature and involves the poor and their associated social consequences: poverty, hunger, unemployment, and health issues. Another is more micro in nature and focuses on how all this money will divert research from many individual diseases. The reasoning runs like this: if $3 billion has been diverted for the HGP and if I am a medical researcher, I will most likely be pulled in the direction of the HGP because its resources are wider and deeper than resources for research on single-cause diseases (for example, Collins's own cystic fibrosis research).

The rebuttal runs like this: the HGP project, if fully successful, will have application in every single-cause disease. Furthermore, it will help to address at least one of the social consequences of poverty, health issues, by pointing to solutions for all health issues, whether for the poor or the rich. In addition, the HGP will help begin the process of establishing human genome libraries for future research, meaning that it will pay certain dividends now and much higher dividends many years from now. By furnishing maps coupled with detailed nucleotide sequencing information, the study of genetic diseases could be advanced much farther than where it is now.[24]

Acquiring the huge volumes of sequencing information promised by the HGP is now well underway. HGP is not an American project but a truly international one, with work ongoing in more than a dozen countries by scores of scientists. By combining computers and mapping, scientists are able to tap into the HGP's databanks that would have taken years to compile had this been a one-team or one-country operation.

For example, the human sequence, if typed out, would require 200 telephone book-sized volumes to get at the human body's 10 million, million cells.[25] These data are now in computers all around the world and will eventually contain not just human genome maps but also maps of thousands of organisms. Of course, these aren't really maps like the ones you might use to go from Charlotte, North Carolina to Nashville, Tennessee. Rather, these are maps of genetic markers that locate the makeup of an organism or a particular feature of that organism.[26] Researchers can then take a sequence, key it into any of a number of HGP web sites for expressed sequence tags (ESTs), and discover what the sequence markers are.[27] So far there are more than 12 million sequences available, including 4.5 million human ones.[28] But this is only the beginning. Once the code has been broken (or, rather, deciphered), many experiments will be performed to give meaning to all of the code.[29] In many ways, reading the code now would be like trying to make sense of this book if all the words were written together, sdrawkcabtubsihtekil.[30]

The experiments and other research will eventually make sense of all this. The human map will be known to any who will take the time to understand it.

It is just this map and where it will lead that concerns many, however. Once this substance and essence of humankind is known, who will protect it, and from whom? On the one hand, scientists like Watson see it as only so much good. "When finally interpreted," he wrote, "the genetic messages encoded within our DNA ... will provide the ultimate answers ... and help us understand [so many diseases] that diminish the lives of so many people."[31]

But not everyone is so altruistic. This information could be used not only in commercially abhorrent ways via genetic screening, but it could also be used by terrorists to determine the most effective means of killing millions of people at one time. Furthermore, if genome libraries are established and records of genetic testing of individuals are kept, this information could be used in various ways that would prove detrimental to either individuals, groups, or both. The potential for harm is tremendous, especially if the sequences are turned over to privately owned banks of genes or used for the patenting of gene probes.[32] Leon Kass's argument that the HGP will only make our lives more complex may resonate more clearly now.

Some scientists, including Robert G. Martin, chief of the microbial genetics section of the Laboratory of Molecular Biology of the National Institute of Diabetes and Digestive and Kidney Diseases, questioned "the urgency it is receiving" and could not "fathom the haste." This sentiment was echoed by the Harvard microbiologist Bernard Davis and scientist Evelyn Keller. The lack of speed bothered others, so one company has marketed InstaGene, a matrix that allows one to speed up isolation while also cutting costs.[33] Most scientists, however, remain sanguine, explaining that the project represents the "encyclopedia of life" and, like that familiar reference tool, it can be used to solve many of today's health problems.[34] The HGP has become at once the great project of our time and "the terrible gift."[35]

It cannot be discounted that much of the widespread acceptance of the HGP in the public (among those who know about it) resides in the fact that the population is now made up largely of aging baby boomers.[36] At first blush this may seem unnecessarily obscure or recondite an explanation, but upon further reflection it makes eminent sense. These aging boomers are the flower children of the late 1960s and early 1970s, the generation of "all you need is love" and anyone over 30 could not be trusted. Now that most are well past 30 and heading toward retirement, life and its ills have caught up with them. The curiosity of science in general, coupled with its can-do attitude, has meshed perfectly with a population suddenly confronted with its own mortality. A perfect marriage has been veritably born, one that many fear has been made without proper thought regarding its long-term consequences.

CONCLUSIONS

The Human Genome Project is a massive undertaking by hundreds of researchers who span the globe. Billions of dollars, both public and private, have been poured optimistically into this venture, all with the hope that what we find at the rainbow's end will be humanity's pot of gold against disease. Only a curmudgeon could argue with this grand and humane plan that could, if ultimately successful, end Alzheimer's disease, cystic fibrosis, cancer, and dozens of other diseases devastating only a very small percentage of the population, and even the commonplace diseases like hypertension that affect almost half the population.

But troubling clouds have already risen on the horizon and the code has not yet been fully analyzed. Gene therapies to make us younger, smarter, faster, and more beautiful are being pushed as hard as cures for fatal diseases, in some cases even more so because the dollars to buy them are ubiquitous. Large gene libraries whose safety and security cannot be guaranteed are vulnerable to anyone who wishes to exploit them. Insurance companies, not known for their altruism, are lined up to tap into these libraries and write policies accordingly. If person A has markers X, Y, or Z, his or her policy might cost thousands of dollars more than those for people without such markers or, worse, not written at all. Universities are rushing into ventures with private companies for commercial genetic enterprises worth hundreds of millions of dollars, compromising at once those doing the research and the universities' very independence. States granting stem cell research privileges are raiding the researcher pool in states that have passed laws prohibiting it. Meanwhile, human dignity has been lost to as many As, Gs,Ts, and Cs as parents pick out, like so many clothes, to adorn their children with various grandiose characteristics.

It's a brave new world and no one knows where it will end or even if it will be habitable. At the dawn of 2005, that end, given the current juxtaposition of so many other variables, does not bode well for our future. Given our past history in this area of knowledge about human makeup (eugenics), and our history with just about any other kind of knowledge, we would be wise to tread slowly and carefully, especially where angels fear our footsteps. Given the enormous egos in science, we would do well to provide more than a little oversight for all of genetic engineering.[37]

7

The Doctor Will See You Now: Genetic Engineering and the Treatment of Diseases

Throughout this book medical breakthroughs have been mentioned in connection with various subjects within the large rubric of genetic engineering. In this chapter, however, these medical marvels (and some not so marvelous) will be examined in greater detail. The examination lends itself to five broad areas: general genetic medical issues (both actual and philosophical); stem cell developments (including the adult stem cell versus embryonic stem cells); so-called spare parts genetic research; specific diseases and treatments (including one-gene diseases, diseases that have multigenetic and multifactorial causes, and gene therapies and screening); and, for lack of a better term, enhancement therapies (including all reproductive genetic issues, such as designing your own baby). Earlier chapters have mentioned some of these. It is the intent of this chapter to treat these areas and many others more specifically while highlighting the controversies they raise.

GENERAL GENETIC MEDICAL ISSUES

Genes have been suspected of carrying diseases for some time. Indeed, Nobel laureate Paul Berg argues that, "I start with the premise that all human disease is genetic. You can sit here for an hour, and you can't get me to conclude that any disease that you can think of is not genetic."[1] Treatments of genetic diseases have been around for a lot longer than one might think. Given that Watson and Crick's famous double helix discovery occurred in 1953, it is only natural that one would place the beginning of genetic treatment of diseases very late in the twentieth century. But that would not be entirely correct. What makes physicians and scientists hopeful about genetic treatments is the history of medicine to date and the process inherent in genetic treatments that have already proven successful in other contexts.

Take, for example, the physician in 1922 who was faced with the untenable fact that his 12-year-old patient would die from diabetes.[2] He took a chance on a new drug made from the pancreas of a cow (a kind of early pharming, if you will). The child improved almost immediately, saved from a certain death. The drug's researchers, two Canadians, Frederick Banting and Charles Best, had finally discovered which part of the body's mechanism regulated glucose. They called their substance isletin. It is now more widely known as insulin.

Most of medicine's breakthroughs have resulted from just this sort of cutting-edge risk-taking. Vaccines for various deadly diseases such as whooping cough, diphtheria, measles, and polio, for example, are now commonplace but had once been shot-in-the-dark treatments. Organ transplants have followed the same sort of cutting-edge, even almost mad-scientist approach. Who would have thought 60 years ago that heart surgery would become routine or at least so commonly practiced that even small community hospitals (at least in America) provide access to it? The same year Watson and Crick made their famous discovery, the first successful kidney transplant occurred. Cataract surgery is done as an outpatient procedure today, while laser vision is being used almost as often as getting fitted for glasses; indeed, it may one day replace them.

Proponents of genetic treatment of disease often point to these once unheard-of techniques as reasons for pushing the envelope (according to some) on genetic engineering. If what could not be thought of 50 years ago is now commonplace, isn't it reasonable, they argue, to expect that in another 50 years all these genetic treatments will become equally commonplace? Moreover, they point to some palpable successes by new, genetic-based companies like Aventis, Genentech, Amgen, and Centocor, along with major drug giants like Eli Lilly and Schering Corporation.[3] These companies and their recombinant products treat large populations suffering from a variety of diseases, such as hepatitis B, cancer, arthritis, heart disease, strokes, and diabetes. For example, in the case of cerebrovascular disease, there are over 110 heritable disorders, 175 with genetic loci, and over 2,000 unique mutations predisposing one to stroke.[4] What is not yet known is how many of these will respond to genetic treatments, or how well.

But the optimism isn't limited to drugs and gene insertion alone. Scientists are excited about the prospects of genetic manipulation, for example, extracting omega-3 fatty acids (normally available only in fish or fish extract) from any animal that has been genetically altered to produce them; this may yield a treatment for heart disease.[5]

Researcher Gregory Pence, himself sympathetic to genetic manipulation, warns that there are at least 10 ways we go wrong in thinking about genetic manipulation:

1. Conceptualizing issues in simplistic oppositions.
2. Creating Olympian standards for new options.
3. Distrusting the choices of ordinary people.
4. Confusing distributive with evaluative questions (questions of enhancements, such as perfect babies, are separate from distributing life-saving drugs).
5. Demonizing new inventions in medicine as "new technologies."
6. Allowing sensational cases to skew our thinking.
7. Ignoring the reality of mixed motives in people.
8. Failing to note the opportunity cost of doing nothing.
9. Ignoring the role of money in bioethical decision making.
10. Failing to advocate positive changes in bioethics.[6]

Pence believes these fundamental mistakes force us into tight ethical and moral corners and prevent us from seeking the benefits, or at least blind us to the benefits, while enlarging the risks. If we avoid these pitfalls, he argues, we will be able to make better decisions, both morally and ethically, about bioethical issues as we save larger and larger populations from avoidable (or, at least, highly treatable) diseases. Besides, we use some genetic manipulations already, and have for years. Take the case of PKU testing, which has been done on every infant (in nearly all industrialized countries) for more than three decades. No one views this with alarm (it is a genetic testing for markers), so why would anyone view testing high-risk populations for even more deadly diseases such as sickle cell anemia or thalassemia?[7]

Genetic manipulation is considered by some to be nothing more than an extended development in medical technology. If people can be saved not only from debilitating diseases but also from inherited ones, why not take great risks, they argue. Even genetic enhancements are considered by these proponents as making much good sense. If a student can boost her score on the GRE by taking several classes—and no one views this with alarm—then why not make a few genetic changes earlier on (say, in childhood) and boost her I.Q. by a standard deviation or more?[8] If this seems too objectionable, argue these proponents, then what about human growth hormones (assuming they can be shown to work), not only for those who suffer from dwarfism but also those who are well below average height for their age?[9]

Many men and women change their appearances throughout their lives to make themselves look better while forestalling the effects of time. A tuck here and a stretch there belie their chronological ages by making them appear much younger. Why not apply genetic manipulation if it can be put to work on the aging process, too? If you could take a pill to remain younger longer, why not take it?[10] How is slowing down the aging process any different from making short people taller, or those who are intellectually average slightly brighter? Isn't this really all the same thing?

Yes and no, say others. Because genetic treatments occur at the molecular level, those treatments can affect a great deal more than we know. For example, the Human Genome Project (discussed in the previous chapter) has revealed a mountain of information. Even when gene sequences are known and correlated with diseases, it will not then be a trivial matter to treat them.[11] Moreover, this new database of knowledge will make it harder and harder to train physicians.[12] Some fear that the adage "A little knowledge is a dangerous thing" may be coming back to haunt us. Faced with a glut of information that cannot be easily assembled and used wisely, some will use all they have or all they have access to. Like the blind man and the elephant, they may be seeing only a part of the whole and miss that whole for the part, with disastrous consequences. Others point to the lethargy with which we have used what we do know.

Rewriting genetic code via gene insertion is looked upon by some as following in "Frankenstein's footsteps."[13] Others, like Marc Lappé, raise the specter of eugenics, arguing that this is another way of attempting to gain genetic control but with certain failure: "[E]ven such genetic control [genetic manipulation] would not assure control over the actual product since nongenetic forces would undoubtedly supervene to shift the direction and end point of development far from the desired prototype."[14] Bear in mind that eugenics, as was pointed out in Chapter 2, has an ugly history. Not only did more than 25,000 sterilizations occur in this country, but that number was exceeded in Great Britain and increased tenfold in Nazi Germany.[15] In other words, we can only plan so far even with the new techniques and the new technology. Besides, how many of us, when faced with the decision, will turn over our future or the future of our children to some laboratory technique that may or may not work as advertised? We may solve one problem (say, reduce strokes in highly susceptible populations) but may unleash something else, perhaps much worse, in the near or distant future.

Lappé goes on to point out that our "manifest dilemmas [rest] on the false premise that a common genetic or chromosomal constitution damns the carrier to a common fate." Lappé, who is himself a cancer immunologist, is quick to point out that "Despite the dramatic consistency with which a few scattered tumors are linked to genes, the problem of cancer as a whole is much more than a problem in genetics."[16] Bernard Rollin argues that scientists and others often make the mistake of thinking that "the truth of moral statements is unrelated to the truth of scientific statements," or that they are "unrelated to the truth ... about how the enterprise of testing true scientific judgments should develop."[17] Once we understand these interrelationships we can avoid many of the ethical dilemmas that confront us now. It is important, say he and others, that we not allow this to become a discussion among experts alone because the ethical and moral issues are so far-reaching.

Gene insertion also bothers some people simply because we do not fully understand what happens following it. "Geneware that 'ships' with the human body," writes one expert, "encodes complex responses to different possible environments, and as genes change, so do the responses. You aren't the baby your parents brought into the world, at least not in the majority. You load new software or forget to update the old and find yourself broken."[18] If this is true of original gene software, what of inserted, foreign gene software?

We do not know as much as we think we do or as much as we think we see in genetic study. For example, we have known the molecular basis of sickle cell anemia for more than 40 years. When first discovered, it was thought that the disease would be easily treated in about a decade. The fact remains that treatment for sickle cell anemia has not changed much in those 40+ years.[19]

Then there is the problem of private funding and applications for patents on what some would call life. In 1991 the U.S. Patent and Trademark Office had 4,000 expressed sequence tag (EST) patents on hand. By 1996 there were 350,000 and by 2000, more than half a million, and this is only the U.S. market.[20] But funding bedevils the research process, too. For example, although there are 16 million people in the United States afflicted with diabetes, we are spending only about $20 per patient (per year). Contrast that with, say, those afflicted with AIDS or the AIDS virus. Only 775,000 known AIDS patients are in the United States, yet we spend just over $2,400 per patient (per year).[21] Some would argue that this is skewed. Recall the chapter on the Human Genome Project. The issue is the same with medical treatments and genetic manipulation: are we spending too much on what may be less essential than more pressing diseases that could possibly be cured? It depends on whose ox is being gored, as the saying goes. A diabetes sufferer might say we're spending too little; an AIDS sufferer, not nearly enough.

The problem may also lie in our ignorance, both in the near and short term. For example, we cannot know what these treatments will do to patients 20 years from now. If we make all the treatments available and subject to insurance claims, we have no idea what the near-term or future costs will be. Furthermore, we have not begun to train physicians to know when to recommend and when to withhold such treatments. As soon as one type of cancer has been cured through gene insertion, it's unlikely that any physician will ever be able to escape its recommendation for the treatments of all types of cancer, regardless of whether the treatment is appropriate.

Then there are those unforeseen, unintended consequences. For example, patient John Moore, whose spleen cells had been labeled by a physician as a "gold mine" owing to their application for diseases of the spleen, lost his bid to protect his own cells. The law protected everyone—science, physicians, researchers—but not the donor.[22] Whether the law will remain this way or be overturned (not unlike frozen embryos) is unclear.

Current blunders are already apparent, say opponents, as the case of Jesse Gelsinger (mentioned in Chapter 3). That case has just been settled for a little over one million dollars.[23] Martin Cline (mentioned in a previous chapter) performed the first gene insertion experiment on two women, one an Italian, the other Jewish, for beta thalassemia, an inherited blood disease. Although his university, UCLA, withheld permission for his experiments, he performed them anyway in the women's respective countries.[24] Thus, the first such experiment created a hailstorm of ethical and moral issues. Questions have been raised recently about genetically manipulated mosquitoes designed to be disease-free, and a ban on fish that glow.[25] Fears are that we are moving faster in the area of technology than we possibly can in ethics, assessment, or regulation. Some argue that by the time these issues catch up, we may have done incontrovertible harm.

Once all of this information is stored, some say we will have a new fear, biopiracy.[26] All of this data, pointing to diseases along with their predictive factors, and the actual screening results of individuals will prove a formidable temptation to those who are looking for a quick dollar.[27] Still others, like the Nazis, may want to create the perfect race but this time from the inside out and with scientific know-how to avoid previous mistakes. Just how much will that information be worth to mad, and not so mad, men?

In the case of genetic companies like Celera and Genentech, it's safe to say we know how much. For example, Craig Venter and his company, Celera, invested everything on this idea of radical breakthrough in diseases via genetic manipulation. Researchers believe that some 1,600 diseases can be treated in some manner via genetic manipulation, though they may disagree about the strength of the connection between a given gene and a given disease. But the strength of the gene-connection case being made has sent Celera stock's market value up 600 percent.[28] On the one hand, this is good news for free enterprise; on the other, that kind of rise (and subsequent fall) is more than enough to tempt individuals to make it work, and that has opponents, both in and out of the field, worried.

Furthermore, these companies have already begun the move from strict disease cure to enhancement through what has been called cosmetic pharmacology or prosthetic gene therapy.[29] For many, it is one thing to call for the cure of cancer but quite another to sound the clarion for stronger jawlines or more aquiline noses. Are we headed for a series of television reality shows based on "radical genetic enhancements" or "choose your perfect baby"? Some experts in the field worry that we are not only headed there but may be already well down that road. For too many opponents, we are not only playing God but thinking his thoughts for him.[30]

Many of the diseases marked for genetic cure are as much behavioral as they are genetically inclined (for example, some top killers are heart disease, cancer, cerebrovascular disease, accidental injury, chronic lung disease, and diabetes). For these, genetic manipulation may be a moot point.[31] Although new research is being done in what is known as behavioral genetics, it is still quite new and even more controversial than genetic engineering. We are not our genes, reminds one scientist, but we may well become them if we allow ourselves to be manipulated by them or by those who wish to manipulate them.[32] Overreliance on genetic manipulation for enhancement, disease control, and cure will simply ensure that too much control over too much of our lives will be placed in the hands of too few. We do not have to go back many years in our history to see that even experts, well-meaning or not, can lead us in the wrong direction. For a full generation, Mendelian genetics was outlawed in Russia while genocide was hotly pursued in Germany, both with the full backing of the science of its day.[33]

Privacy issues abound in this new era of genetic manipulation. For example, genetic testing can be done for a variety of reasons: preimplantation diagnosis, prenatal diagnosis, newborn screening, carrier testing, diagnostic analysis, confirmatory analysis, presymptomatic or predictive testing, and susceptibility testing.[34] While each of these types of testing has a legitimate medical basis (though some may be less persuasive than others), each one also represents a very serious and potentially damaging privacy issue. While scientists may argue over various tests, all agree that predictive and susceptibility testing open up new and more challenging ethical and moral issues. Guaranteeing that these results will never be seen by anyone but for whom they are intended is simply impossible in this age of computer hackers. While we might not want to forbid testing for this reason alone, it is certainly a strong enough reason to want to make it particularly clear to those who are thinking of undergoing genetic testing for any reason.

The ethical implications are enormous, according to some. The problems include discrimination based on genetic diagnostic tests; negative practices from therapeutic abortions; positive and negative eugenic practices including in vitro techniques (either diagnostic or from advanced technologies); the pharming of human embryos for organ transplantation; immoral uses of embryos in experimentation or commercialization; and negative practices in animal pharming.[35] While both sides are fully aware of these problems, these issues do not appear to be fully addressed beyond naming them. The events of September 11, 2001 have added bioterrorism to this formidable list of recombinant DNA concerns.[36] Given the calculus of our world during the war on terrorism, we cannot let current events cloud right thinking on this vital issue, nor the proper response to those who will use recombinant DNA for exclusively evil purposes.

MEDICAL USES OF STEM CELLS: THE CURRENT DEBATE

Stem cell usage, especially embryonic stem cells, continues to stir up debate. The debate focuses on whether researchers should be allowed unlimited access to and use of embryonic stems cells (cells that have not yet been triggered to become certain kinds of tissues) versus using adult stem cells (cells with a specific usage that are found in bone marrow and organs such as the spleen, liver, nasal tissue, brain fat, and umbilical cord blood). While adult stems cells are available throughout the human population and can be extracted relatively easily (though not painlessly), embryonic stem cells must be taken from human embryos that are between five and seven days old. The extraction process is fatal to the embryo.[37] The argument focuses on the promise of embryonic stem cells. Some say they hold the potential for curing 100 million patients with various diseases. Others not only dispute this number but argue that embryonic stem research is entirely unnecessary. Opponents argue that adult stem cell research shows far better promise while embryonic stem cell successes have been overhyped. Furthermore, these opponents argue that research on donated umbilical cords is just as effective and has had great success in treating a number of diseases, most notably Gunther's disease, Hunter and Hurler syndromes, and acute lymphocytic leukemia.[38]

Almost every reader will have heard something about this debate due to the very public campaign made by actor Christopher Reeves before his tragic death in 2004. Mr. Reeves had suffered a horseback riding accident that left him paralyzed from the neck down. His efforts and the efforts of his foundation pushed this young research into the forefront of dinner table conversation. Former President Ronald Reagan's son, Ron, also made the debate even more public during the 2004 presidential elections. The young Mr. Reagan came out in favor of the democratic candidate, John Kerry, even though his father had governed the country for two very successful terms as a staunch, conservative Republican.[39] Senator Kerry had come out during the campaign for a much stronger and far-reaching program than the Bush administration had (or had proposed) so far. Doubtless Mr. Reagan's departure from his father's conservatism had something to do with Reagan Sr.'s Alzheimer's disease.

Research to date on embryonic stems cells is very sketchy. Animal experimentation, for example, has shown very mixed results—nothing conclusive. Embryonic stem cell research on diabetes showed promising results early in 2001, but subsequent research has shown that the insulin-producing cells generated from embryonic stem cells produced only one-fiftieth of the normal amount. In the same context, a 2004 study revealed the alarming tendency of embryonic stem cells not only to develop insulin-producing cells but also to develop deadly tumors in mice.[40] On the one hand, this is hardly surprising, given the infancy of the research. On the other hand, the scant successes would tend to warrant slowing down embryonic stem cell research.

Herein lies the debate. Both Presidents William Clinton and George W. Bush have pursued the same methodical process, as was pointed out in an earlier chapter. The U.S. policy on embryonic stem cells is more permissive than the policy of 23 other countries, among them Germany, Spain, France, and Canada. Ten countries have embryonic stem cell policies much more permissive than the U.S. policy, the United Kingdom, Japan, China and Israel among them; yet it is the U.S policy that gets most of the criticism.[41] While use of embryonic stems cells has had only mixed results, use of adult stem cells has been more promising. Adult stem cells have been used to successfully treat Parkinson's disease, cartilage defects, blindness, lupus, multiple sclerosis, rheumatoid arthritis, severe combined immunodeficiency disease (SCID), and cancers such as various leukemias, solid tumors, neuroblastomas, non-Hodgkin's lymphoma, renal cell carcinoma, and certain childhood neurological diseases.[42] Some new research indicates that there may even be therapy for improving your memory.[43] Clinical trials using adult stem cells have been successful in treating severe heart attacks, as well. Nevertheless, as is true in most of the history of genetic engineering, promise often exceeds delivery. For example, recently all of the optimistic information regarding multiple sclerosis and genetic engineering took a black eye when the second patient on drug therapy for the disease was confirmed to have a rare and deadly infection.[44]

So why the debate? Why not pursue adult stem cell research vigorously while slowing down embryonic stem cell research? In part, that is what is being done, at least in the United States under George W. Bush. President Bush's advisor, Dr. Leon Kass, himself a respected biologist and physician, has been the chief proponent of this argument and has been making it in compelling fashion for the last 30 years. But others argue for a much more permissive policy that would allow research on both fronts simultaneously. While these proponents understand that the policy would offend some by destroying embryos, they contend that the possibility of saving many others, even millions, outweighs any other consideration. This almost pure utilitarian approach frightens some while alarming others.[45] Moreover, there is now a chance to slice the ethical hairs even thinner, thanks to inheritable genetic modification and embryo selection. The latter involves preimplantation genetic diagnosis (PGD), which allows the discarding of the embryo before implantation; this means that one could conceivably discard an army of embryos before deciding on the utility of one.[46] The debate isn't likely to end, however, as the country remains very nearly evenly divided on the morality of saving some at the expense of others, even when those "others" are not yet fully developed embryos. The debate sharpens its focus in favor of current policy when the argument becomes one of creating embryos only in order to save other lives.[47]

The stem cell debate may become moot if a recent study proves correct. A study published in January, 2005 argues that current embryonic stem cells, at least those for which there is federal funding for research, may have been contaminated by animal molecules.[48] All 22 colonies or lines of human cells that had been approved by the government had been cultured with the aid of an animal feeder that expresses an acidic sugar, N-glycolyneuraminic acid (Neu5Gu). The acid is found in nonhuman animal cells but is not found naturally in human cells. According to the study, this acid has now been expressed in the human cells in such a way as to trigger the human cells to attack the animal ones as foreign bodies. The study argues that the lines or colonies are now useless for human treatments and new lines are required. To obtain the latter would mean destroying embryos in order to create new lines. This new bit of evidence will doubtless have to be studied carefully to confirm these early suspicions. Meanwhile, it has added new fuel to the stem cell fire.

Finally, California's recent initiative approving $300 million annually for stem cell research has increased the intensity of this debate. The amount makes the state of California a leader not only among U.S. states but also among countries like Sweden and Singapore that have made stem cell research a national priority.[49] The U.S. federal government has set aside only $24 million annually for stem cell research. Many states fear that California's initiative (Proposition 71, passed during the 2004 elections) will cause a brain drain from their own states. Massachusetts, for example, is now seeking to finance stem cell research and is urging its lawmakers to pass laws that would make stem cell research legal and encouraged. Others states, such as Florida, are seeking one-upmanship. In Florida, where the President's brother is governor, a private group is campaigning for a state measure that would provide $1–2 billion for embryonic stem cell research. The proposed financial bonanza from Proposition 71 has been scaled back according to a recent report. The report argues that Californians have "unrealistically optimistic expectations" about potential financial gains.[50]

Even the headline-grabbing techniques that created Dolly (discussed in Chapter 5) are coming under reinvestigation. For example, a recent study found that the cloning procedure (somatic cell nuclear transfer) used for Dolly and a host of other cloned animals is so inefficient that it raises questions about the procedure with respect to animals, not to mention humans. That process required 277 attempts, using somatic cells that are easily obtainable from adult animals. However, the extreme inefficiencies raise questions about using animals as bioreactors for pharming (i.e., using farm animals as testing laboratories for drugs and organ donors).[51] Not only are the inefficiencies a concern, but some critics contend that somatic cell nuclear transfers also result in the premature death of surrogates, calling into question both animal and human cloning.[52] Others argue that the whole cloning project (including the Human Genome Project) creates a "have versus have-not" technology that may cause governmental powers to refuse the technology to certain people or entities, resulting in a kind of free-for-all over the technology, effectively ending all research of this type.[53]

THE BIOTECHNOLOGY OF SPARE PARTS

If all of this isn't enough, there is now the issue of creating spare parts for humans, like so many pieces of machinery at your local automotive repair shop. Researchers are already thinking about human fabrication and have begun to create human heart valves, breasts, ears, blood vessels, pancreases, cartilage, noses, and other body parts.[54] The two men most responsible for advancing the field, Robert Langer, Professor of Chemical and Biomedical Engineering at Massachusetts Institute of Technology, and Dr. Joseph P. Vacanti, argue that "The idea is to make organs, rather than simply to move them."[55] Some argue that this is only the next logical step beyond using animal parts for human replacement parts. Others find it morally and ethically uncomfortable to create body stores, if you will, where spare parts for humans could be purchased when original parts have failed or have aged beyond maximum strength and utility.

The technology for this process is well underway. Along with it, of course, is its commercialization. Unless the technology changes, acquiring these organs would require destroying embryos or growing them for this purpose alone. Obviously, not everyone is in agreement with a process that would require a harvest of would-be humans in service to those who have already fully matured and possibly abused what they were born with. While some contend that we have an obligation to use body products in a manner that benefits everyone (except, of course, those embryos garnered into service), others see it as the exploitation of the living.[56] The issue has again raised the question of to what extent any society will tolerate playing God with human lives.

SPECIFIC GENE THERAPIES: WHERE ARE WE NOW?

Chief among the hopes for stem cell research treatment is the debilitating and eventually fatal disease cystic fibrosis (CF). CF has what scientists call a "clear and consistent pattern of recessive inheritance," allowing researchers to look for the specific CF gene that causes the trouble. The process is known as restriction fragment length polymorphism analysis and allows scientists to pinpoint the problem on the DNA chain.[57] When a marker is located near a specific gene, the chances are high that the marker and the gene will be inherited together. What has troubled scientists, however, is that though they have come so very close to unraveling this mystery they are no closer to a cure.

Recent studies indicate that a CF patient's own stem cells might aid in his or her treatment.[58] But the study is very new, more work must be done, and the therapy, if successful, will not really offer a cure but will aid in the treatment of patients with this dreadful disease. CF is a very difficult disease to treat because a breakdown in the lungs' cells obstruct the air passages. This happens at such a relatively rapid rate—and there is nothing to retard its progress—that treatments to date have prolonged patients' lives but have not cured any. One researcher points out that while stem cells may become useful in this search for a cure, the extension of CF patients' lives has come from "good nutrition, aggressive antibiotic use, meticulous physiotherapy, and increased understanding of the disease."[59] It is important to remember, however, that this disease, (in which about 1 in 25 Americans carries a mutation in a CF gene) can be easily spotted in very reliable, low-cost testing.[60] Is it time to require more general testing of larger portions of the population?

This is a question that sparks much controversy, as indicated earlier. While genetic testing may save lives, as some argue, it can also make many others live lives of not so quiet desperation. Testing positive does not necessarily mean a person will get the disease. In this case, the right information could be too much information, much too much. Calls for more legislation that would make testing mandatory have not gotten much, if any, traction.[61] The failures or near-failures of prior testing, such as in the early cases of phenylketonuria screenings (PKU, mentioned earlier; which has now been largely perfected) and sickle cell anemia raise important yet unresolved ethical questions. While there is some effort to make genetic screening more widely acceptable, each new false positive sets the program rapidly rolling in a backward direction. How individuals will take the news (and this does not include how outsiders such as insurers will use such information) is key to this controversy.[62] Meanwhile, the promise for an asthma-like inhalant treatment for chronic obstructive pulmonary disease (COPD) by 2010 continues; for now, it is scarcely more than that—a promise.

Cancer also tends to have a predilection toward stem cell treatments.[63] Scientists are enthusiastic because they have been able to determine that as cells go from healthy to cancerous, certain very definite changes occur. The cells de-differentiate, or no longer specialize.[64] They also appear to be immortal, meaning they can divide without limit as well as incorporate other behaviors that lead to a patient's death. By examining these cells in vitro, scientists are coming to understand how to undo the damage caused by cancerous cells. More study, it is argued, will lead to an eventual breakthrough that will allow some genetic alteration that will either retard or stop the progression of cancer.[65] But scientists also agree that many behavioral factors militate against any potential treatments, however optimistic. For example, even if advancements led us to a full recovery from some cancers via stem cells, researchers are in agreement that it would have little lasting effect if the behaviors that trigger the disease were not also held in check.

Successful DNA treatments of Crohn's disease, an inflammation of the lining of the small intestine, have also been encouraging. Using DNA technology, scientists have been able to send a signal using an antinsense molecule to inhibit the inflammation process.[66]

The use of recombinant DNA technology has also discovered single-gene effects in schizophrenia, manic depression, and Alzheimer's disease.[67] Further work has identified the unique, blood-based molecular signature of schizophrenia and bipolar diseases.[68] One company, ChondroGene of Toronto, used its Sentinel Principle to arrive at the blood test. The company is also working on similar blood tests for cancer, cardiovascular diseases, and arthritis. Furthermore, Duke University researchers have located the first direct evidence in mice of the role of serotonin in the brain. It is hoped that this will also lead to better future treatments of psychiatric disorders.[69]

Inherited retinoblastoma, a kind of cancer of the eye, affects only about 1 in 200 children in the United States, but scientists have been able to treat it using DNA technology. Heretofore, the disease, which grossly enlarges the eye, required treatment via removal of the eyeball. By discovering that inherited retinoblastoma was caused by a mutated DNA segment, scientists now believe that if they catch it early enough they can replace the missing or mutated gene.[70] A certain form of colon cancer, hereditary nonpolyposis colon cancer (or Lynch syndrome), is also responding well to DNA therapy. Once the offending gene could be marked, scientists have been able to retard or stop its growth by suppressing it with a tumor-suppressing gene.[71] Researchers are equally excited about the discovery of RNA interference, a tool that allows them to turn genes on and off. Early research indicates that it will have far-reaching effects on the elderly, who have poor to failing eyesight, allowing them to turn off the gene carrying degenerative eyesight. And that's only the beginning.[72] If successful, the RNA interference will have long-term implications.

Not everything on the stem cell/gene implant front has to do with disease. One organization, the Center for Neural Engineering at the University of Southern California, is working on a silicon chip implant that apes the hippocampus of the brain, the area responsible for creating memories.[73] It's unclear now how, or even whether, the chip will work in a live test but, so far, the research is very optimistic. Recreating long-term memories in patients with brain damage, or simply creating better ones, may be possible soon, according to researchers at the Center.

Another nonmedical use is that of transgenic microbes in landmine detection.[74] Certain bacteria sense what is called trinitrotoluene (TNT, by its more familiar appellation) as well as other kinds of explosives. Since these landmines leak their explosives, the idea is to aerially dust suspected fields with the microbes, allow them to work, and then shine ultraviolet light over the field at night. Each mine would then light up its location. The test has been conducted in my own home state, South Carolina, over a 300-meter area. The microbes located all five hidden landmines. The uses of this technology are obvious and would save many lives in war-ravaged areas.

About 1 in every 3,500 male infants in the United States is born without a functioning gene for an important muscle protein called dystrophin.[75] Lack of this protein leads to muscular dystrophy. The most common form is Duchenne muscular dystrophy, where the muscles gradually wear away. Death is virtually certain for these patients by their early twenties. Muscular dystrophy is very difficult to isolate, but in 1991 scientists discovered a technology developed at Vical, Inc., San Diego, that allowed for the injection of dystrophin into the thigh muscles of mice. The research is very promising but less than 3 percent of the mice injected actually produced the necessary dystrophin protein to prevent muscular dystrophy.

For more than a decade scientists have known that there were certain inherited markers, or mutations, for breast cancer, BCRA-1 and BCRA-2.[76] In 1997, famed gene hunter Mark Skolnick of The University of Utah Medical Center disambiguated the long arm of chromosome 17 and found the mutant DNA that now goes by the name of BCRA-1.[77] That discovery meant BCRA-2's days were numbered. That same year, David Goldgar (Utah), Michael Stratton (Institute of Cancer Research in Sutton, Surrey, England), and Doug Easton (also from the Institute of Cancer Research) found the second culprit on the long arm of chromosome 13. Genetic testing is encouraged among women to see if they have these markers. BCRA-2, for example, may increase (albeit only slightly) a woman's chance for ovarian cancer, too, and so if a woman tests positive she may want to consider having her ovaries removed as a preventive measure. At any rate, the presence of BCRA-1 and BCRA-2 plays a role in invasive, aggressive cancers and signals physicians to treat the matter most seriously.

For other types of breast cancer, however, knowing what to do isn't so easily understood. What we do know is the prevalence of these diseases and this helps to explain why researchers push so hard to know more. One in eight women in the United States will get breast or ovarian cancer, meaning that the likelihood that you will know someone who will fall victim of the disease, and know them well as a friend, mother, sister, or aunt, is very high.[78] But not all breast cancers are alike.

For example, one type of breast cancer, ductal carcinoma in situ (DCIS) is puzzling.[79] DCIS originates in the milk ducts in a woman's breasts. In some cases, this cancer never metastasizes or becomes any more of a problem. In other cases, it appears and spreads so rapidly that treatment is often a foregone conclusion when discovered. When this cancer is discovered in a patient, physicians often do not know whether to treat it aggressively or wait and see what develops. With new genetic testing, however, physicians can now link DCIS, BCRA-1, and BCRA-2 together. Now women with a family history can be tested and, if the markers are discovered, can decide whether to have more frequent mammograms and other tests or have a mastectomy. Certainly the presence of two markers would signal a call for more aggressive treatment.

Treatment of these and other diseases with stem cells or similar gene therapies is simple, really. All gene therapies involve the transplant concept, though without organs.[80] The trouble is, the technologies are very, very complex. Though few scientists thought we would be as far along as we are by now, the inherent complexity of these therapies and the sloppiness of some treatments have yielded few successes, even among diseases thought to be readily treatable.[81] It is the speed at which we arrived at stem cell and other gene therapies, coupled with their low success rate, that has become both the victory and defeat of genetic engineering. In many ways, genetic engineering promises so much while delivering so little.

Thalassemia, the blood disease that affects many African Americans, for example, seemed at one time to be on the brink of cure. As has been pointed out in previous chapters, success with this disease has eluded scientists but for no apparent reason. It is almost as if we can see the fruit of success on the tree; we just cannot figure out how to get it off the tree and to market, so to speak. It should work, yet for some inscrutable, maddening reason it does not. In most cases, all we have learned (as, for example, in Martin Cline's research mentioned in an earlier chapter) is that a particular gene therapy doesn't work.[82] This is, of course, quite helpful in pointing scientists in another direction but it has not advanced the process to the ultimate goal of cure.

All sorts of strategies with RNA, DNA, and retroviruses have been tried. Most, to date, have failed. Some, as in the case of insulin with its subsequent tumor-inducing outcomes (mentioned above), have worsened the condition. Luckily, trials have been slow and animal research has contraindicated the treatments for humans. It is important to stress this point, given that many criticized the government for what they view as being overly conservative or foot-dragging. At this stage of what we know, it would be irresponsible to rush the process. In many cases, it would appear that the diseases in question are simply more complex than anyone imagined.

The push for cures is understandable. Lesch-Nyhan syndrome is a case in point. It affects young children and often leaves them ranting, vomiting, and thrashing about with uncontrollable biting behaviors, as if they were in the throes of demonic possession.[83] Obviously, no physician wants to treat such children unsuccessfully; hence, the rush for a cure.

Yet we need not focus on one disease alone or limit ourselves to diseases of children to understand why some push so hard for success at any cost. How many of us want to face—either ourselves or with our loved ones—the slow withdrawal from reality that is the fate of Alzheimer's patients? It is just this rush against time, fear, or disease that leads to false hopes, profound disappointments, and even deadly treatments. Furthermore, because this research is so very costly we cannot deny that the rush to successful, marketable treatment also figures into the overly optimistic hopes, or in treatments that are sometimes worse than the diseases' outcomes.[84]

Although the process that led to famed researcher W. French Anderson and the birth of gene therapy proved a most difficult one, the road ahead is not free of potholes.[85] While genetic technologies have created a medical autobahn, its ethical and moral potholes remain unfilled.[86] We have also seen that no failure is insignificant. For example, work done with Down's syndrome has given scientists clues about directions to take on Alzheimer's, even though the work on Down's syndrome per se has been futile to date. Even limiting gene therapies to those that do not cure immediately but over time (as, for example, the case of injections for cardiovascular diseases) provides enormous savings in money and a considerable reduction in pain.[87]

To argue that we must stop would be to concede not only obvious failure but consign people with these diseases (and those who will surely later get them) to lives without hope.

Odd as it may be to say, stopping this research would also stop work on diseases that have yet to make an appearance today. How can that be? It has been estimated that within each person there arise at least 22 new germline mutations.[88] While not every mutation is bad, as we have said, researchers believe that some of these will surely lead to other untreatable diseases for which we will forever be searching for silver bullets. We are also discovering that the step from knock-out mice and transgenic or cloned animals such as Dolly to genetic germline engineering in humans is not a small one.[89] Ninety-five to 99 percent of all engineered embryos, for example, are lethally damaged. A more reliable and safer process must surely be found. Perhaps a more methodical approach would allow more time to assimilate this ever-changing scenery.

Slowing down the process or proceeding more cautiously and more carefully might allow for a better resolution of the very intractable moral and ethical issues. To ignore or minimize those issues would, however, lead to certain disaster. Clearly, these are difficult questions. On the one hand, they may lead to great cures or, at the very least, the prolongation of life. They may also lead to terrible tragedies along the way. Is this a price we are willing to pay? It's a question we cannot take too lightly.

ENHANCEMENT THERAPIES

Perhaps we are willing to make the case that we can heal some people at the expense of others. And most would argue that providing for couples who cannot have children to be able to one day is surely an unqualified good. But what of the so-called enhancement therapies that lead to so-called designer babies? Designer babies are the as-yet unborn whose DNA is tweaked to improve strength, weight, height, and speed, not unlike a sports combine. In other words, designer babies are chosen with an eye to creating better babies. Are we morally and ethically ready for this?

We had better be. Gene therapies and gene implantation make it possible for many of these traits to be hand-picked ahead of time. Moreover, given our technological expertise and our openness to reproductive rights, couples will not merely choose; they will be able to take a peek ahead of time and, if unsatisfied, delete and start again. It's a brave new world after all!

Some are not ready for these prospects but they also know it won't be easy to forestall them. Writes one researcher, "[T]he political challenge for those opposed to designer babies is daunting. Success will require widespread recognition that eugenic engineering would both trample human rights and undermine inviolable commitments to social and economic equality. Eventually, a permanent global ban on human germline manipulation and reproductive cloning will be necessary."[90] Note that the old bugbear from Chapter 2, eugenic engineering, appears once more in this conversation. Is it possible for humans to begin a neoeugenic approach that will produce only good? The odds appear against it, even at these early stages. "People will cease to place a premium value on their 'natural' attributes," writes another researcher, "over those which have been acquired synthetically."[91] Leon Kass, President Bush's genetic engineering point man, puts it even more starkly:

Man is the watershed that divides the world of the familiar into those things which belong to nature and those things which are made by men. *To lay one's hand on human generation is to make a major step toward making man himself simply another of the man-made things.* Thus, human nature becomes simply the last part of nature which is to succumb to the modern technological project, a project which has already turned the rest of nature into a raw material at human disposal.[92]

Apparently some fear the coming future could well be one long "extreme makeover" television reality show. Unfortunately, early research and commercial products indicate that this fear is neither misplaced nor overdramatized. Have we ushered into the world quality control of the unborn?

At the same time, much of this research has led to discoveries that enable couples who have markers for terrible diseases, such as cystic fibrosis, to abort fetuses that have the same markers and try again.[93] While there are certainly those who would bristle at the idea of abortion, how many parents, knowing their soon-to-be-born son or daughter carried the marker responsible for 70 percent of all cystic fibrosis cases would ignore that datum? Is it more ethical to abort the unborn than to bring it to term?[94] In other words, if so-called enhancement research yields opportunities to prevent tragic, disease-ridden lives, would it be more ethical to prevent them? Or is it more ethical to prevent such research at all? To put it another way, spilt milk is indeed a terrible mess to clean up, and about this no one will disagree. But is the mess so awful that we would consider outlawing milk? While this analogy, like all analogies, is unfair (milk is a staple of life; genetic engineering is not), it does perhaps put in high relief the devil-in-the-details nature of these questions. As we have seen in every chapter of this book, the questions are neither easily nor quickly resolved.

Designer babies are only part of the battle. Plato's parenting dream in the *Republic* appears to be within reach. In the *Republic,* Plato questioned whether parents were in fact the best individuals to raise their children. The emotional attachments often precluded, in his mind, reasonable (i.e., ratiocinative) upbringing. But from Plato's day to this, no one really gave it much thought since parents and children are inextricably linked from the moment of conception until Junior goes to college.

In this context, since 1995 Nobel prize-winning biologist Jean Rostand has envisioned the creation of an artificial womb.[95] Scientists now believe they can create one and will soon be able to develop a child completely outside its mother's womb, from the moment of conception to birth. Should this come to pass, it is only a few steps away from creating more human spare parts in those artificial incubators. Science has always marched to the mantra that it can improve upon nature. The question now is, is this an improvement? Some would argue to the contrary.

Even in fail-safe arrangements, things can go awry. Consider the famous case of surrogate mother Mary Beth Whitehead and the *Baby M* trial in 1987. Ms. Whitehead had signed a contract indicating that she would abstain from cigarettes and alcohol, undergo amniocentesis, and, if a genetic defect was found, she would undergo an abortion if the genetic father required it. Whitehead attempted to gain parental rights and failed; the genetic father's wife adopted the child. The court overturned this ruling and gave the parental rights to the genetic father, proving a well-known adage: if it ends in court, you court disaster.[96]

Consider another even more complicated case, the custody battle for embryos in Tennessee.[97] Seven embryos were frozen for future use when the couple in question filed for divorce. The frozen embryos became the battleground for one of the most unusual court cases to date.[98] If a legal and so-called binding contract does not unravel these issues, as we make them even more complicated with genetic technology, what hope have we to make them any more comprehensible? These may seem far from the subject of genetic engineering but they were far from reality's horizon, too, before they emerged. They may give us some idea of the complicated scenarios that can result when we posit unrestricted genetic engineering.

CONCLUSIONS

What is the average person to make of all this? On the one hand, we have the possibility of uncovering great cures. On the other hand, we have the prospect of choosing whether your children will have blue eyes, be as musically gifted as Mozart, or as athletically talented as Shaquille O'Neal. If gene therapies are ultimately successful, will Congress set the new retirement age at 95 for the next generation?[99] These are not idle questions. Can we come to some agreement about them before they overrun us? For example, would it be fair to say that vaccines or therapies that cure people should take precedence over those that enhance normal abilities or give the average person superior ones, especially if enhancing therapies are likely to be sought by many more people than those in search of cures?

Questions about the process are sure to discomfort for some time to come, and agreement over them would be helpful. We have seen how unreliable and unsafe genetic engineering techniques can be. Can we agree to slow down the process until we can arrive at a process that is more humane and less wasteful? Is it right to create hundreds of thousands of embryos through in vitro fertilization but then never use them—to throw them away or send them to research laboratories?[100] Perhaps an even more important question is whether it is right to have so many created at all.

All such questions converge on one running theme of this chapter: while our technology has progressed rapidly, our ethical and moral thinking about these questions has not kept pace. In many ways it could not. The jump from helping one disease-stricken man dying of treatable heart disease to a disease-stricken clone dying from what we are not sure of is a gigantic one.

Heart disease provides a case in point. In 2001 scientists announced, amid much fanfare, that stem cells extracted from bone marrow and injected into the damaged hearts of mice managed to become special cardiac muscle tissue that the body cannot replace after a heart attack. All geneticists were agog over what they understood to be elementary biology when it comes to stem cells: put them in the right place and the body will do the rest. Now, four years later, there is no evidence that that is possible in humans and so a debate (naturally) is waging over whether it should even be expanded to humans at all.[101] While such failures do not wreck everyone's hopes that some benefit may eventually arise, it does indicate that any promises are not only premature but foolhardy.

It should be clear by now that questions remain that must be answered, if not by this generation then certainly by the next. The longer we put them off, the more intractable they become. We are rightly dazzled at the prospect of curing diseases, alleviating or eliminating pain, or prolonging life. Taken independently or in their pristine forms, nothing could be more awe-inspiring. But our story doesn't end there. Rather, it is more like the well-intentioned man who saves a pedestrian from certain death only to learn later that the same pedestrian murders an innocent person in the park for her purse. The horns of our dilemma could not be more pointed. But that is all the more reason we should resolve them before we find ourselves gored by them.

8

Just the Facts, Ma'am: Genetic Engineering, DNA Evidence, and the Courts

Turn on television any night of the week and you're bound to see some version of *CSI*, the crime scene investigator series featuring our genetic superstar. Whether in Miami, New York, or the original show in Las Vegas, *CSI* agents, looking for all the world like they just stepped off Paris runways, are tracking down criminals with only a scintilla of eyebrow or a few molecules of epidermis, and, of course, our main protagonist. But one need not turn to *CSI*, though it's hard to miss since it's on three or four days a week. Try *Law & Order, Missing, Cold Case,* or any of the spin-offs of *Law & Order* (and here we're talking about 50 percent of the shows on television, or so it would seem) and you cannot miss the central character, our chief protagonist, our main twist from central casting, DNA.

So far in this book we have seen DNA involved in the treatment or cure of diseases; the central figure in enhancement; or the key ingredient in designer babies. Now we find our central figure in courtrooms around the globe under a variety of aliases: DNA profiling, DNA fingerprinting, and DNA testing.[1] It is almost as if we see that familiar double helix of Watson and Crick sitting in the witness box, expatiating freely, not only about how the crime was done, but when, at what time, with what instrument, and by whom ("Miss Scarlet, in the drawing room, with the wrench"). Although most of us groan when we see the letter from our local magistrate or court official requesting our presence for jury duty, DNA, a relative newcomer, is now omnipresent and so very glad to be there.

It hasn't always been that way, however. So how did DNA go from playing bit parts in test tubes to the lead on the boob tube or the silver screen? More than 20 years ago, in 1983, the body of 15-year-old Lynda Mann lay dead on a path in Leicester County, England; she had been brutally raped.[2] Standard forensic evidence turned up nothing. When 15-year-old Dawn Ashworth turned up raped and murdered three years later in a nearby town, local police called in the big guns. Enter Alec Jeffreys of Leicester University, a well-known researcher in the field of human genetic markers. The police eventually charged a man but Jeffreys, using DNA from vaginal swabs, concluded that while the two young girls had been raped by the same man, it was not the one in custody, one Richard Buckland. Buckland had confessed, but Jeffreys's work proved pivotal and Buckland became the first person to be exonerated by DNA profiling.

The tragic story has a somewhat happier ending. Police had taken DNA samples from some 3,600 males, but without a match. Finally, a coworker admitted to having provided DNA samples for Colin Pitchfork. Police arrested Pitchfork, he confessed, and a jury subsequently convicted him. DNA had been used before, of course, especially in paternity suits, but never in a case like this. By 1989, DNA testing of this sort had been used in 85 cases in the United States, and by 1991 every state had forensic DNA testing.[3]

As common as DNA evidence is today, shortly after it began, it nearly shot itself in the foot. Without knowing it, most readers of Tom Wolfe's 1990 novel *Bonfire of the Vanities* are familiar with the watershed case *New York v. Castro*, which cast doubt on DNA procedures or, rather, DNA interpretation.[4] Lifecodes Corp., one of two companies in the 1980s doing DNA testing for the courts, had performed the DNA testing for this case but the interpretation turned out to be rather slipshod. Although this case cast some undeserved doubt in the minds of many about DNA testing, what it really did was establish DNA testing as a means of proving one's guilt or innocence, but only when performed properly. Lifecodes's lax procedures widened the debate on standards for interpreting results, laboratory practices, and band-matching.[5] *Commonwealth of Massachusetts v. Lanigan* and *Commonwealth of Massachusetts v. Breadmore* (1992) also challenged interpretations and process. While the court agreed that DNA testing was a well-established and well-accepted theory, the method by which the particular lab in these cases went about establishing test results was proven to be a generally unaccepted procedure.[6]

If all of this were not enough to create reasonable doubt, one Joyce Gilchrist, an Oklahoma forensic scientist who helped send dozens of convicts to death row, came under the gun for substandard practice just a few years later.[7] Gilchrist had worked for law enforcement for 21 years and won the Civilian Police Employee of the Year award in 1985. Her conviction rates earned her the sobriquet Black Magic. It now turns out that her results may have been influenced by her own biases about each case. Oddly, though many convicts were put behind bars with Gilchrist's help and her DNA results, DNA retesting in her cases is now releasing just as many.

DNA results have also had far-reaching effects outside the courtroom, though still in an identify-the-victim scenario. Rumor had it that Czar Nicolas II of Russia had not actually been executed during the Bolshevik uprising in 1918—he had escaped and substitute bones had been placed in his grave. However, DNA testing of those recently exhumed bones left no doubt as to who was buried in the Czar's grave.[8] Walk into your local Wal-Mart, any mall, or any one of dozens of grocery stores and you're likely to see DNA fingerprinting being offered free for children. This fingerprinting is being done to aid in the hunt for children should they come up missing. But such testing does not always have happy results.

In the late 1970s, Argentina's military suppressed dissidents and more than 9,000 people disappeared.[9] Many children orphaned by these circumstances later found homes with military families who were childless. Eventually, grandmothers began searching for these children and 15 years later, thanks to DNA testing, many were found. By that time the children were teenagers and many were aghast to find that their adopted parents were the very ones who had had their biological parents killed.

Outside the courtroom, DNA is also being used to preserve animals at risk of extinction.[10] Zoos around the country are beginning the process of preserving *Jurassic Park*-like gels, not of T. Rex in amber but of nearly extinct species such as the wombat and the South African blue antelope. The idea is to provide enough genetic coding or profiling to reconstruct endangered animals at a later date, if necessary. Some question this process, calling it crass commercialism because the animals will likely be used for entertainment purposes (in zoos) or scientific sensationalism, because we cannot say with certainty if the animals would be healthy or normal. Others, however, believe the preservation of any species far outweighs all complaints to the contrary.

IS DNA FINGERPRINTING RELIABLE?

Can we be sure that DNA fingerprinting or testing is accurate? Can we be sure that a given criminal is precisely the right one when his or her DNA turns up? How so? Thirteen specific sites in a DNA strand about five feet long identify each of us as, well, us. It works like this:[11]

- Samples of a fluid (blood, semen) or skin tissues, eyelashes, or eyebrows are collected.

- 100 cells are required and the cells are put in a chemical and heated to release the DNA.
- A polymerase chain reaction is induced (a synthetic primer that duplicates the 13 comparison areas millions of times).
- The DNA is zapped with high voltage, separating the 13 DNA sites and making them glow.
- A computer reads the glowing DNA and prints peaks and valleys as well as the markers at each of the 13 sites. The FBI has assigned numbers to each possible configuration, making each unique.
- The process is repeated to make sure it's right. The resulting conclusions are as certain as anything can be; the probability is 1 in one-hundred million that the DNA belongs to only one person in the world.

The process, as described above and perfected by Jeffreys, allows for a quick and easily verifiable way to identify individuals. Though the strand of DNA is five feet long, and though the human body has 3 trillion cells (all except the red blood cells containing DNA), the process isolates the 13 required to distinguish you from me and both of us from every other person, plant, or animal on the planet. A technique known as Southern blotting produces what appear under the microscope to be bar codes not unlike those seen on grocery store products.[12] DNA fingerprinting is like traditional fingerprinting. In the same way that no two people can have the same fingerprints, there are no two individuals who have the same DNA fingerprints (with the possible exception of identical twins).[13] DNA profiling is generally used for serious crimes such as rape or murder, hence its association with television's most popular crime shows.[14] The errors discussed above were not errors in the process so much as they were errors in interpretation, pointing out once more that while science can claim near-certitude for its conclusions, humans involved in the process can quickly reduce that certitude to reasonable doubt.[15]

Most forensic profiling works on the basis of one of three processes: Variable Number Tandem Repeat (VNTR), Short Tandem Repeat (STR), and Sample Sequence Variation (SSV).[16] VNTRs are lengths of DNA made up of 15 to 35 base pairs in variable repeats, yielding 500 to 10,000 base pairs located at a specific site on a certain chromosome. The number and size is inherited in a Mendelian manner. The sample can come from blood (with white blood cells), tissue, saliva, or semen. The DNA is digested with a restriction endonuclease enzyme that cuts the DNA into four to six base sequences.

These sequences are placed in a gel and are negatively charged. The results are the bar code-like bands mentioned above. These are inherited as markers for that individual. Because VNTRs require large amounts of undegraded DNA, polymerase chain reaction (PCR) is used for degraded smaller amounts of DNA. PCR is useful for STRs. The last process is generally used for identifications based on protein differences. VNTR has been used to identify Gulf War victims and the victims of the explosion of TWA Flight 800 over Long Island in 1996.[17]

In another, earlier case (though not involving murder), a Ghanaian boy was refused entry into Great Britain in 1984 because immigration officials suspected the woman with him was not his mother. The father was not present and the mother, moreover, could not with certainty name the father. Tests revealed that the probability that either the mother or father would by chance posses all 61 DNA markers of similarity to be 7×10^{-22}. Within the 61 DNA markers of similarity were 25 maternal-specific fragments and the probability was 2×10^{-5}. If the supposed mother was in fact the boy's aunt, the probability that she would share maternal-specific similarities with her sister was 6×10^{-6}. The boy was thus allowed residence with his mother in Great Britain.[18]

Probably no more significant instance of failure of evidentiary interpretation (though not DNA interpretation) could be cited than the familiar O. J. Simpson case. DNA testing in that case set the odds that someone other than the famous footballer committed the crime at 1 in 170 million people.[19] While the science proved certain, the prosecuting attorneys in the case made a serious blunder by allowing the defendant to try on a wet leather glove over a dry one. That, plus the doubt cast upon the way blood samples were collected and processed in the lab, provided enough reasonable doubt to set the defendant free. Given that DNA testing set the odds remarkably high, the verdict proved most controversial. For example, if the blood had been tested at five points on the DNA, and that matched what had been found at the crime scene, then odds that the perpetrator was not O. J. would be 1 in 10,000,000,000, or more than the number of people on planet Earth.[20] As Alec Jeffreys put it,

You would have to look for one part in a million million million, million before you would find one pair with the same genetic fingerprint, and with a world population of only five billion it can be categorically said that genetic fingerprint is individually specific and that any pattern, excepting identical twins, does not belong to anyone on the face of this planet who ever has been or ever will be.[21]

Can DNA profiling uncover mutations that are predictors of violent crime? In 1965 cytogeneticist Patricia Jacobs published studies of the chromosomes of 197 men who had been committed to the Carstairs Hospital for the criminally insane in Scotland.[22] The study showed a strong predilection for criminal activity among men with schizophrenia, a mental disease that begins in adolescence and worsens with age.

When the news reached the United States, the story was sensationalized. Some considered it sound eugenics to abort the offspring of men with the XYY chromosome. XYY is a rare chromosomal disorder in males that generally results in violent and antisocial behavior. It is sometimes referred to as the 47XYY. Others believed the behavior of schizophrenic criminals could not be controlled, that it was inevitable. Still others thought that if there was any chance that children might have this gene passed on to them, their would-be parents should be sterilized. Much of the hysteria had to do with the timing of the study, the 1960s. At that time there was no gene industry, no drugs, no treatments, nothing but the furrowed brows of physicians telling frantic parents to be watchful.[23] By the mid-1970s, however, Jonathan Beckwith of Harvard and Jonathan King of MIT had conducted another study that demonstrated the flaws in the XYY studies up to that time. Beckwith and King focused on the ethics of telling parents they had a child with the XYY chromosome and how this could label a young boy from the moment of his birth and throughout his life.

The matter remains highly controversial. It does appear that the presence of an extra Y chromosome along with gabella and long face may be predictive of future trouble. Throw into the mix learning disabilities and early behavioral problems, and criminal proclivities might be unavoidable. That study, along with continued research, has led to complex ethical and moral issues. Should families with the extra Y (known as fragile X) chromosome be warned? Should testing for these families be mandatory? Like all DNA testing, the presence does not mean destiny. Testing will not be complete until the Human Genome Project fully maps the 20,000 to 25,000 human genes. Should we discover that certain mutations, combined with other factors, strongly predispose individuals to violent behavior, what would be the proper ethical response? Will we eventually see cases resolved in favor of innocence because "My genes made me do it?"

Genetic databases like the Human Genome Project will prove critical to the success of DNA fingerprinting. These banks will provide valuable data in catching criminals or exonerating the innocent. Several European countries are developing them, too, with Great Britain as the first of the EU countries (1995), the Netherlands and Austria next (1997), Germany the following year, and Finland and Norway in 1999.[24] Each country already governs the uses of its genetic databanks. The process whereby data are added, subtracted, searched, and the like results in very different databases. One wonders how useful these may be in the future if some agreed-upon criteria do not become universal.

The issue of crime and genetics has a more recent controversy, as well. In 1992, a National Institutes of Health conference on Genetic Factors in Crime had to be cancelled amid a flurry of vituperation and animosity. The NIH had launched an initiative to test inner-city children for biological markers such as 5-hydroxtryptamine, low levels of which are supposed to make individuals more violent. Although the NIH had planned to test all inner-city children, because at that time most of them were African American some people alleged that the conference promoted racism.[25]

This is but one of the many issues that make the matter of DNA profiling no easy ethical matter. Privacy issues are also important. Since DNA can come from virtually any particle of a human body (an eyelash, an eyebrow, spittle, blood from a cut, and dozens of others), many have stressed the importance of securing prior consent.[26] As mentioned earlier, DNA fingerprinting of children (in case they are ever abducted) is also controversial. Although there is great benefit to having this DNA information on file, many observers fear it could later be used against those tested by looking for more than markers that simply identify them (e.g., looking for markers of various diseases).

THE GAY GENE

Even more controversial than this would be the revelation that a given child possessed the so-called gay gene and was predisposed to homosexuality.[27] It should be pointed out, however, that the research on this gene is very controversial and highly doubtful.[28] It has not been replicated and the only researchers to claim its authenticity are themselves gay. Other research has been done that shows, for example, that the anterior hypothalamus (a place in the brain) is in gay men indistinguishable from that found in women.[29] Another anatomical difference was discovered in 1991: the interstitial nucleus of the anterior hypothalamus is smaller in the brains of gay men than in the brains of heterosexual men. All that can be said about the research, however, is that it shows some very small differences. These findings could just as easily be the result of expressing one's homosexuality as a cause of it. Furthermore, there is no evidence to date suggesting that men are gay apart from behavioral and environmental factors.[30]

Studies to find the gay gene are ongoing. One such study awarded by NIH to Alan R. Sanders, a psychiatrist at Evanston Northwestern Healthcare Institute (outside of Chicago) is embroiled in controversy. Sanders will essentially try to replicate and improve upon Dean Hamer's work in the five-year grant to study 1,000 pairs of homosexual brothers. Hamer's research was the first to make a claim for the gay gene. After his results were published in *Science* magazine, National Public Radio and national magazines (such as *Time* and *Newsweek*) rushed to press with stories about the gay gene. They were premature. The controversy stems from some believing that biology cannot dictate sexual orientation (in other words, there is no gay gene), while others worry it will launch a new round of testing that could provoke parents to abort any child, male or female, who tests positive for a putative gay gene.[31]

DNA BANKS AND TERRORISM

In addition to databanks with DNA sequences from all over the country, especially from young children, there is also the potential for companies to test employees for DNA identification in this age of terrorism (such databanks would have dramatically sped up the identification process at the World Trade towers). There is also routine DNA testing now performed by the military for easier, quicker identification of remains.[32] On the surface this seems only sensible, but imagine recruits (or employees) revealing data that marks them for heart disease or cancer. Can we be certain such evidence will never be used against them, never figure into promotions (in or out of the military), or never impact downsizing or any other similar decision? Most do not see how such information would not inevitably be used to impact an individual in a negative manner. Is this enough to halt the genetic engineering process and the panoply of benefits from DNA profiling? Again, most say no.

CONCLUSIONS

DNA profiling will continue and will also expand and enlarge its value. Whether providing a clue (usually the only reliable clue) to a serious crime or tracking down a lost child, DNA profiling will continue to augment its usefulness to our society and its well-being. New techniques promise to make it even more reliable than it is currently (though this is hard to imagine), while also improving the very area of profiling that needs improvement—interpretation.

DNA profiling isn't likely to reduce crime, though by its very function of narrowing down the suspect list to a handful or fewer, it may well give pause to the smarter class of criminal (oxymoronic though that may be). The power of DNA's predictability is certainly underscored by the many so-called cold cases that have gone unresolved for years but are now routinely being solved by new DNA evidence. In 2004 the U.S. House of Representatives attempted to pass a bill that would allow all federal inmates access to postconviction DNA tests that could exonerate them, but opposition killed it.[33] The bill will likely resurface in 2005.

DNA profiling may make it possible to reduce capital punishment to a handful of cases. In other words, when DNA profiling is used in capital crimes, its finger-pointing ability is a mathematical certainty. Given that turn of affairs, capital punishment may eventually be restricted to only those crimes where

DNA profiling can positively identify the perpetrator (and so would reduce its use). This would provide verdicts with such mathematical certitude that convictions will no longer be made worrying over possible convictions of the innocent. Either way, DNA profiling could bring an end to the arguments about capital punishment that have been going on for the last 100 years. Its use will be certain and its frequency rare. Only the fewest ardent opponents on both sides will be unhappy.

What isn't likely to change, however, is the continuing controversy over databanks that hold DNA profiling results. The knowledge provided by such banks has the potential to violate the privacy of individuals while maligning their well-being if the data fell into the wrong hands. How this problem will be resolved is anyone's guess. So far, efforts to provide fail-safe mechanisms to prevent would-be hackers or other unauthorized users or uses have failed. Given that any computerized system can be hacked into, it is unlikely that such a system will be forthcoming in the foreseeable future.

9
Endgame: Genetic Engineering, Future Trends, Current Recommendations

The last eight chapters have been compiled in an attempt to provide a broad and general background to the debate over genetic engineering. As we have seen, it is far more than mere cloning, though that subject will always be the screaming headline-grabber. The so-called Xeroxing of humans will forever overshadow any other discussion on this topic, but it's important to know that genetic engineering encompasses many other subjects, such as the genetic modification of foods, modification of human genes in an effort to find cures for diseases, or even modifications to produce human enhancements of some kind. To talk about genetic engineering is to talk about all of these topics. To better understand the debate requires study and careful attention to all the component parts that make up this very complex subject. Alexander Pope, that master of the two-liner, once wrote,

> What dire offenses from am'rous Causes springs,
> What mighty Contests rise from trivial Things.[1]

DNA is hardly trivial, as it turns out, but given this infinitesimally small element found in chromosomes one is hard-pressed to think of a tinier object from which has sprung so much woeful ink, so much optimistic press.

This final chapter will focus on four areas in an effort to sum up what we know so far and where we might be headed in the future. These four areas provide the basis for continued research and further discussion. If we refuse to treat these areas, it is likely we will usher in yet another round of continued controversy and endless bickering. If we will address ourselves to these large areas of concern, perhaps we will begin to find common ground to solve our common problems. These areas are eugenics (redux), scientific values and scientific ethics, new discoveries and new problems, and playing God.

EUGENICS (REDUX)

As was shown at the close of Chapter 2, much was written about eugenics and its implementation in the early part of the last century. Stellar figures rushed to the forefront of this movement and called upon leaders and government to take charge of creating, if not a best race, certainly a better one than we had.

The advent of enhancement technologies, along with cloning and other features of genetic engineering, raises the specter of eugenics once again, and in a far more serious manner than before because we now have the microtechnology to make it happen, and quickly. Well-intentioned authors talk of so-called positive eugenics and using genetic engineering to improve our genetic pool. That movement is well afoot and is being defended once again by those at the forefront of science.

"What was misused in the eugenics movement," writes one scientist, "was not science; it was lack of science, it was ignorance. . . . I say that with strong conviction because I think it is not warranted to say that science has been misused."[2] On its face, this sounds either too optimistic, too naïve, or both.

Furthermore, eugenics came to us via origins that were "broadly progressive, and often in the left-wing [of] thinking."[3] In a red state/blue state nation this is important.[4] All too often we are dismissive of conservative, largely right-wing views and much of our thinking in this matter has been influenced not only by the media but also members of the professoriate. We are far more likely to embrace a broadly progressive view simply because we ourselves wish to be thought of as progressive and forward-thinking. We are less likely to turn from a progressive viewpoint than we are to turn to it. After all, it was none other than Plato who recommended in his *Republic* that the so-called ruling classes be bred responsibly.[5] When *soi-disant* intellectuals recommend to us a path to take, we find ourselves irresistibly drawn to it, if for no other reason than we wish to be included with the open-minded. But one can be so open-minded, the Southern writer Flannery O'Connor warned, that their brains fall out. In the case of genetic engineering, we cannot afford to allow this to happen.

"I am the child of a Nazi survivor," writes Jeanne Spellman of the Women's Occupational Health Center at Columbia University,

and many millions of others suffered very directly and convincingly the results of policies based on dubious genetic traits. . . . While I don't see gas ovens being built in the United States, I do see that many of the premises which led to that terrible time are again beginning to surface here.[6]

It isn't that genetic engineering will necessarily lead to another holocaust of horrific proportions, but that ideas circulating about it lean in that direction and so must be carefully and thoroughly vetted. It should be enough to cause us to ask very particular and very precise questions and to view the answers with great care and attention. By not examining these questions and calling science to account, we run the risk of accepting the unacceptable, if for no other reason than we have painted ourselves into that corner.[7]

In all fairness to the eugenics side of the debate, however, at least one scholar is making new claims for a what he calls liberal eugenics that would allow for the full or nearly full implementation of genetic enhancement.[8] In what must be ranked as one of the world's greatest understatements, this scholar admits that "Hitler ... [has] made eugenics an unpopular idea."[9] Indeed, and for no other reason than that alone, we must forever be wary of it, regardless of its sophistication and urbanity.

Have we learned anything from history, from our earlier attempts at progressive eugenics? Can we undertake the betterment of the human race without also demeaning it? How we define "better" will largely affect whether we improve society or only one part of it. To talk of human betterment is one thing, to hold up one part of humanity over all others is quite another. It is not yet clear that we have learned that important lesson.

Healing and breeding need to remain distinct within genetic engineering.[10] Any movement that relieves human suffering caused by disease should be hotly pursued if it does not come at the expense of others. If, however, we intend to link better breeding with healing, we might lose one while promoting the other. We have, without a doubt, entered a brave, new world. It remains to be seen whether we will make of this new world a dream or a nightmare. The resulting outcome may well hinge on the speed with which we undertake our pursuit. If we see knowledge and its acquisition as an unquestionable good, its pursuit will most likely be at breakneck speed and at the expense of ethical matters. If, on the other hand, we see it as having a potential for harm we may well pursue it with all due process. Therapeutic purposes should always outweigh cosmetic ones.[11]

Even in the pursuit of therapeutic purposes, we must continually ask ourselves if the elimination of all pain is not only a good, but also a worthy, goal. Surely it would seem so, but as Shattuck points out, pursuit of all knowledge for its own sake is bound to leave pain in its wake.[12] We must remember that eugenics proved to be the centerpiece of a governmental reform that gave us the modern world's greatest horror, the Holocaust. It would be silly to argue that any future eugenics effort will lead down this horrific road again; it would be equally foolish, however, to contend that any modified eugenics, or so-called beneficial eugenics, will never make a wrong turn.

The Faustian man who would sell his soul for knowledge is the man of excess.[13] Such men are like gods who bring fire from heaven to burn their benefactors, not warm them. Well-intentioned does not mean doing well and it's hard to see how eugenics, however it is resurrected, could ever become an innocuous good. In every age, whenever eugenics emerged some lives were ranked above others. Unless one gets to choose his or her category, it's not likely that this new version of eugenics involving recombinant DNA will be any different. It's important to remember that human nature has not changed. We are the same people, behaviorally, we were 2,000 years ago. All that has changed is the speed at which we are able to make changes. This ability is as much a hindrance as it is an advantage.

SCIENTIFIC VALUES, SCIENTIFIC ETHICS

What science values is not the same thing as scientific ethics. Indeed, the latter is only now getting the attention it deserves. In many ways, what science values remains a conflict of interest with what is ethically right.[14] The rise of the biotechs has put extreme pressure on scientists to produce. Some have rightly challenged this as being one reason why scientists cannot be judge and jury of their actions. As one scientific magazine put it, "If an investigator is comparing Drug A with Drug B and also owns a large amount of stock in the company that makes Drug A, he will prefer to find that Drug A is better than Drug B."[15] But the conflicts aren't always financial, either; sometimes they are egomaniacal in nature.[16] Oftentimes the success of scientists pushes their own formidable egos out from view. The general public may well misattribute to altruistic motives what really springs from nothing more than small-minded, personal interest. Scientists are as heir to the foibles of flesh as the rest of us and oftentimes scientific pursuits are made for little reason beyond making the pursuer better thought of than before.

Financial conflicts of interest remain the most serious problem in the research to date, however. Although Harvard University stipulates that any researcher who owns more than $20,000 in publicly traded stock cannot serve as the principal investigator on a research grant funded by the same company, it may still not be enough, according to some observers.[17] In many ways, this may be the most stringent of rules to date but it is not terribly binding. Even a co-investigator or an investigator involved only marginally can still dramatically influence the outcome.[18]

Recall the celebrated case of John Moore, mentioned in an earlier chapter. Moore received cancer treatment and recovered. During his treatment it was discovered that his spleen produced large amounts of proteins such as interferon and interleukin, which are known to aid immune systems. Researchers patented the stem cell line from his spleen and sold it to a Swiss pharmaceutical company for $15 million. The company went on to make billions. When Moore discovered this, he sued because there had been no disclosure or financial return for his cell line, which was nicknamed the Mo cell line.[19] Moore did not succeed, as the court decided that to find in his favor would interfere with the progress of science, though the court did grant him some financial compensation. "So there is unquestionably a tendency," writes one scientist, "to ignore or minimize dangers growing out of scientific activity."[20] While this may be too strong, certainly science is slow to acknowledge iatrogenic outcomes (outcomes inadvertently created by a medical procedure). Iatrogenic outcomes can just as easily be bad as they can be good.

Moore's is not the only case that could be highlighted. By synecdoche, it is a part that serves for the whole. Where will impartial observers be found? Courts are too clogged with frivolous lawsuits to hear and decide these cases in a timely manner. Citizens are too ill-informed to be able to sit in on research efforts and inform panels of scientists on the best approaches. Scientists themselves appear to be far too inextricably tied to outcomes to be impartial observers. As Sheldon Krimsky observes, "This train of thought leads us to the following query: Given the incentives for following the experimental results wherever they lead, what evidence is there that the private investment in academic research plays any role in shaping the outcome of a study?"[21] Early evidence indicates that it could very much influence the outcome by slow negative or nonpositive results, or influence it by suppressing particularly damaging materials. The race to map the human genome is a case in point here, as much of the evidence that came forward was not fully ventilated until later. The race became as important as the outcome.

One reason for the hypercharged debate over values within genetic engineering has been that scientists generally do not consider technical activity as value-laden. Perhaps this is because scientists pride themselves on being able to separate themselves from the conclusions they come to. They are, in fact, no more (or less) capable of that than the rest of us.

It isn't because they have not been warned. Indeed, 20 years ago some scientists were unfurling the red flag. William Lowrance wrote that "Technical activity must be considered value-laden in two senses: technical people's social values and value perceptions affect their research and service; and that work, in turn, affects the value-situation of others in the public."[22] Lowrance considers research a value-laden exercise, as it certainly is. He goes on to argue that technical experts make crucial decisions for and on behalf of the public. Add to this mix the literally tens of millions of dollars at stake and you have a very flammable combination. "Serious trouble arises," Lowrance goes on to point out, "when the distinction between facts and values is blurred or not recognized, or when disputants engage in mislabeling."[23] And so it has, especially in the case of stem cell research and the debate over embryonic and adult stem cells. As was pointed out in Chapters 3 and 7, though great success has been seen in the use of adult stems cells versus embryonic stem cells, and though the rejection factor with embryonic stem cells has not yet been conquered, embryonic stem cells continue to grab most, if not all, the headlines, not to mention findings. Given that most people in search of a stem cell solution are desperate (either for themselves or for their loved ones), clear, honest, and methodical discussion should be mandatory. For example, it is now thought that while adult stem cells have been proven successful in a number of diseases, use of embryonic stem cells will "almost certainly never be an effective treatment [especially in the case of Alzheimer's] because ... it is a whole-brain disease rather than a cellular disorder like Parkinson's."[24] Many of the great hopes for diseases treated by embryonic stem cells fall into this multifactorial category, making the latter useless in the treatment of such diseases. Embryonic stem cells may prove useless because replacing one gene in a mutifactorial problem changes nothing about the behavior and other factors that influence conditions for the disease.

Science informs our cultural outlook whether we want it to or not. It raises our awareness on issues that the general public cannot be expected to debate on its own, given the level of technical expertise required to understand the terms of the debate. Furthermore, science should (but does not always) provide important analyses of certain technological changes. It challenges society with its advance of technological changes that may force us to accept tragic commitments to the consequences of avoidable outcomes. We cannot disengage politics from the onward, inexorable march of science any more than we can stop politicizing any other social remedy. If this arm of the scientific debate becomes more politicized than it has already, we face serious ethical issues from which there may not be an easy escape. Science has a professional responsibility to make certain its contributions to this debate are as factual and as evidenced-based as possible.[25] Just because science discovers or uncovers something does not make its subsequent interpretation of that discovery factual. Lewis's warning, mentioned in Chapter 3, is an instructive one: "The sciences bring to the facts the philosophy they claim to derive from them."[26]

Because an inserted gene becomes part of the new genome, future generations may be at risk. There may be no theoretical reason to prohibit gene implantation because of the risk that whatever is inserted will be passed down to future generations, virtually unchanged. But should we ever allow any treatment that creates an inherited change capable of being passed down to future generations, however good for the public benefit immediately, to go forward without regard to it future unanticipated consequences?[27] Evidence so far does not indicate that a good gene inserted into a stem cell line will always produce good genes any more than do those we are born with.[28] Furthermore, early cloning successes, including Dolly, indicate that there may be an inherent obsolescence built into the process, since many clones have experienced either a short life span or multiple illnesses and health difficulties far beyond what the naturally born endure.

The potential of genetic engineering to place enormous powers into the hands of a few people is great. It is clear that scientists will become gatekeepers of this knowledge. The question then becomes who will watch the gatekeepers, who will watch the watchdogs?[29] Until the creation of Dolly, human cloning fell into the same category as science fiction. Since her creation, however, it has entered into the realm of near-certitude, though no human cloning has yet been accomplished beyond the blastocyst stage (roughly 100 cells).

Some point to current bans on human cloning as sufficient for now. But these so-called bans are clearly not enough to prohibit human cloning if that is what we intend for them to do. Even if these bans will prohibit human cloning in a certain country (and some would argue that they do not), there still remains plenty of off-shore human cloning freedom for it to take place at will, as well as more than enough private funding to see it to fruition. Clearly, there remains room for the creation of some regulatory agency to oversee all research for what some have called "short-term optimism and long-term risk" and the overuse of scientific discovery.[30] At present, given the great divide between groups who want human cloning to succeed and those who do not, there does not appear to be a way to create such an agency that can regulate it sufficiently to satisfy the warring factions. Calls for trusting both human rationality and logic to steer us to a safe haven seem naive.[31]

This says nothing about the scientific community's response to those within their own ranks who oppose human cloning. Take, for example, David Prentice, cofounder of Do No Harm: The Coalition of Americans for Research Ethics. Prentice has taken an opposition stance on human cloning and has been quoted widely by Senator Sam Brownback and Representative Dave Weldon in testifying before Congress against human cloning. His critics have been quick to avoid his rigorous science and instead pillory him by making sideswipe claims, such as that he uses tobacco industry schemes to "devalu[e] scientific information."[32] The point isn't to elevate one side over the other but to demonstrate that the debate has already degenerated when one side resorts to innuendo and *argumentum ad hominem* in an effort to win the day.

Is genetic technology all that different from other kinds of medical and scientific technology? Is the implantation of genes much different, if at all, from the ingestion of a pill? Both genetic and pharmaceutical medicine involve locating diseases and isolating them by replacing bad genes with good, using drugs to eliminate or alleviate symptoms or enhance one's biological inheritance. But, as Leon Kass observed,

[D]espite ... obvious similarities, genetic technology is also decisively different. When fully developed, it will wield two powers not shared by ordinary medical practice. Medicine treats only existing individuals, and it treats them only remedially, seeking to correct deviations from a more or less stable norm of health. Genetic engineering, by contrast, will, first of all, deliberately make changes that are transmissible into succeeding generations and may even alter in advance specific future individuals through direct "germ-line" or embryo interventions. Secondly, genetic engineering may be able, through so-called genetic enhancement, to create new human capacities and hence new norms of health and fitness.[33]

Moreover, famed MIT researcher Steven Pinker reminds us that enhancement is not asking parents if they would like to have stronger, faster, smarter, and healthier children. Who would ever say no? Rather, is it more like asking them if they would like to undergo an extremely painful, traumatic, and expensive procedure that might give them a slightly more talented child or a deformed one.[34] When put in realistic terms, all the controversy and debate seem moot.

Kass, the George W. Bush administration's point man for bioethics, goes on to point out that such gerrymandering of genes to create a new paradigm or new human capacities removes individual freedom while quashing human dignity. He argues that our fears about genetic engineering are engendered by its challenges to our dignity and humanity—disregarding the one and reducing the other. For others it is the combining of human dignity with money or selling the most recent genetic discovery to the highest bidder.[35] Our refusal to bring genetic engineering under "intellectual, spiritual, moral and political rule" allows it to control us, not the other way around.[36] Furthermore, "[E]ven when a genetic flaw causes diseases, it doesn't automatically mean that it can be [successfully] treated by replacing the defective or missing protein with its biotechnologically created equivalent."[37] The most familiar examples of this are, on the one hand, the failure of cystic fibrosis to respond to genetic transplantation and, on the other, the gene implantation in diabetic mice that results in cancerous tumors. Finally, Kass warns that the discussion of genetic engineering falls to simple pragmatism: does it work, how much will it cost, will it do detectable bodily harm, and so on. This leads to an overreliance on a simple pro or con list with a premeditated desire for the pro side to win.

Still others argue that we are dangerously close to hubris, or arrogance, as we claim for ourselves a right once known only to God or nature alone.[38] What, these proponents go on to ask, about our genetic privacy? Before genetic engineering, there was no need for it. With its advent, we now find that we require laws to protect even the most microscopic of human parts.[39] Others posit more doomsday scenarios. What about ethnic weapons that target identified populations while sparing other populations, and biochemical warfare? What about those headless mice mentioned in an earlier chapter that raise the specter of headless humans created solely for organ donation?[40] Could there be a more awful vision than the Nazi-like butchering of humanoids set aside for the benefit of others? It may sound like science fiction today, but recombinant DNA has not only made these scenarios possible, it has also made them more probable than ever before.[41] And what about diversity of the human species, should cloning ever become commonplace? Bernard Davis observes,

Molecular genetics may also contribute indirectly to the recognition of genetic diversity by promoting acceptance of evolutionary principles. For all genetic diversity has arisen through the mechanisms of evolution—and while in humans the genetic differences are more readily demonstrable for physical and biochemical than for behavioral traits, the latter are subject to the same evolutionary rules, which predict broad diversity.... What an inefficient and dull society we would have if the human species were a giant clone, instead of having a population with diverse talents, and a unique personality in each new baby![42]

Furthermore, as new information arrives we will rush to put it to use. We need to remember, as one researcher has put it, that "Miracles of modern medicine may be, when performed, but curses to the recipients."[43] This was particularly true in the recent case of young children undergoing experimental gene therapy for immune deficiencies.[44] These young children have developed leukemia after their retroviral gene transplant into bone marrow stem cells. This is almost surely the result of the activation of a cancer-producing gene used to transfer the therapeutic genes into the cell. In other words, the cure is worse than the disease, or at least equally fatal.

Testing will come to mind as the Grail of safety, but even there the debate is everpresent. Too much eager funding, some argue, makes testing (regardless of how much) suspect. After all, we have too many instances of questionable drugs getting to market (e.g., Vioxx and Celebrex, both now considered key agents in heart attacks or strokes among those who have taken them repeatedly). There was also Jesse Geslinger, the young man whose genetic therapy resulted in his immediate death. Even what some have called scientist hubris (self-testing, such as that done by Drs. Kenneth Murray and Charles Weissmann) remains suspect.[45] Murray ingested lambda phage to prove it did not remain in a human system. Weissmann twice ingested *E. coli* to prove it did not last long in the intestines since the available "slots" are already occupied and a stronger, more adaptable invader would be required to remove what's there naturally. Some scientists view such testing as grandstanding. Other see it as interesting, even compelling, but believe more study would be required in order to make any resulting drugs a widely available procedure.

If we cannot control genetic engineering in its infancy, how will we do so when it reaches maturity? For example, we are rapidly closing in on intelligence—how it is passed down and on which gene. As soon as we know that, the market will be flooded with requests to the Brainiac Company or Einstein, Inc.[46] The whole notion of genetic enhancement means that we are allowing some people to define what is normal human functioning.[47]

The 65-year old Yale psychiatrist and neurosurgeon Eugene Redmond, who operated on XO47, a green vervet monkey of slightly above-average intelligence, is farthest from most people's idea of a Dr. Jekyll. But after he implanted 3 million human brain cells into the cranium of that monkey, some think differently.[48] If he is right and the brain cells grow as predicted, Redmond may have found a possible solution for Parkinson's disease. If he is wrong, Redmond has created another in a long line of chimeras, some of which were only internally odd, others both internally odd and externally frightening. These and other unanswered questions continue to bedevil the field of genetic engineering.

It is these unanswered questions that have caused the Bush administration to slow down the process and hold what opponents have called "a national seminar."[49] "In enjoying the benefits of biotechnology, we need to hold fast to an account of the human being, seen not in material or mechanistic or medical terms but in psychic and moral and spiritual ones. [W]e need to see the human person in more than therapeutic terms."[50] Republican leaders in the House may, however, be warming to another idea on stem cell research. It is important to bear in mind that the Bush administration outlawed only federal funding for research on new stem cell lines, leaving private funding as open as ever. Recently, however, House Speaker J. Dennis Hastert agreed to schedule a vote on stem cell research (the kind the Bush administration objects to) after some Democrats threatened to withhold votes on a budget resolution. During the interregnum, Senate majority leader Bill Frist, a car-diologist, went on the Senate floor and announced his support for federal funding of new embryonic stem cell research. His announcement in August 2005 is in direct opposition to both the Bush administration's stance and his own only a year ago. Whether Frist has had a change of heart, or only a heated desire to be president, remains to be seen. Regardless, it marks a new escalation in the stem cell research debate. Some observers are calling it a "vote swap of epic proportions."[51] This could cause a crack in the Republican stance against stem cell research, leaving no modulated voice opposed to open and unrestricted stem cell research. If so, how will we prevent becoming the tools of our tools, as Thoreau warned?

To worsen matters for the Bush administration, Korean scientists have had substantial success with stem cell research. Korean researchers report that they have found a highly efficient method for producing embryos through cloning. Once cloned, stem cells can be extracted from the embryos.[52] The star research team, however, had been under investigation for ethical violations in stem cell research in 2004.[53] The news has, however, put a great deal of pressure on the Bush administration. The promise of cures ranging from spinal cord injuries to breast cancer holds much hope for literally millions of sufferers worldwide. The Bush administration has roundly criticized the Korean news and any research that kills life in order to save it.[54] While lawmakers have promised a vote on stem cell research, Bush promised a veto if it violates his previously articulated principles. Meanwhile, the Korean researchers have promised to open up a stem cell bank by the end of 2005 to speed up growing replacement tissues to treat diseases.[55]

Others, however, argue to the contrary about new health norms. Proponents of genetic engineering ask whether the creation of new norms of health and fitness is inherently immoral or unethical. Others (quite a number of others) say no, and many of these are scientists who stand to gain much either from the research, the subsequent investments, or both.[56] Perhaps a new norm is needed, they argue, because the old no longer serves or no longer answers the relevant questions. We are on the brink, say these proponents, of great healing and increasing longevity. All that's required is a little more money and a little more research.[57]

Aren't we already facing new norms of health and fitness via conventional, nongenetic medicine? Fifty years ago, life expectancy hovered around age 60. Today, it has surpassed 70 and is rapidly approaching four score. The retirement rule of 65 years of age in most companies was rescinded more than three decades ago as Americans began to live longer, healthier lives. And what, say these opponents, could be more dehumanizing, more violating to human dignity than a disfiguring disease such as Parkinson's, a debilitating one such as Alzheimer's, or a disabling one such as cancer? Moreover, what could give greater joy than to provide genetically related children to sterile or gay couples?[58] Since all are prime candidates for the genetic engineer's refinement, why not let him or her do her refining?

For example, individuals who have Li-Fraumeni syndrome inherit a defective form of the p53 gene.[59] Most people have two normal copies of the gene, but those who do not have a 50–50 chance of developing deadly cancer before 30 years of age. The risk reaches 90 percent by the time they are in their sixties. Testing of all family members allows for extended lives and early treatments. This is not the only gene responsible for cancer, either. Widespread testing and full-scale research on stem cells is the only solution, say proponents of genetic engineering.

We have always had wars and will always have them. If there is a way to make wars less deadly, or deadly for combatants only, why not do so? That goes double for any medical procedure. All such procedures have risks, from small to great, but risks all the same. What we have discovered over the history of medical science is that practice and repeated use has perfected infant, ham-handed procedures.[60]

Moreover, say these proponents of genetic engineering, there is risk assessment that leads us to choose the right kind of risk assessments over worse ones.[61] Even with its utilitarian overtones, assessments that benefit large numbers of the general public will more often than not be chosen over those for whom the same benefit aids only small cadres of individuals. Great benefits for greater numbers will always trump small disasters, at least in the beginning. Too many small disasters and class action suits bring a halt to scientific progress, however it's defined. Risk assessments in genetic engineering are, however, quite complicated to make. Not only are there the obvious risks affecting individuals, but there are also larger ones affecting the environment, epidemic outbreaks, and the release of pathogens, to mention only a few.[62] Gauging the right risks at the right time could mean great progress or ultimate failure.

Some argue that politics now plays too much a role in discussions about genetic engineering and that politics have no place in science. But science and politics have been in sharp tension for centuries.[63] As mentioned in an earlier chapter, the Bush administration holds virtually the same views on stem cell research that the Clinton administration held (and, oddly, there was then little talk of science-politics tensions). Genetic engineering got a great deal of play in the last election, largely because the Kerry-Edwards camp made more claims for the future than the Bush administration was willing to go on record about. Moreover, the science–politics debate goes far beyond mere genetic engineering, covering ecological concerns, global warming, and more.

PLAYING GOD

Given the power genetic engineering possesses to change lives both today and in the next millennium, it will come as no shock that its opponents have accused scientists of playing God. The phrase is a metaphor but it is also descriptive. Since science can now bring its recipients the ability to change much about their lives, including the potential to change their looks, intelligence, and possibly even their personalities, not much else would be left for God to do. Scientists have themselves beckoned their colleagues provocatively, summoning them with "Come, let us play God."[64]

If we are to allow science to play God, we must be certain that science has not only the power to create but also an equally creative power to correct. "Science has taught us the hard lesson that human beings and their Earth are not 'the center of the universe,' but it is now placing in human hands the powers and responsibilities to remake decisions we formerly left to God."[65]

Of course, this is not the first time humans have tried via science to control their environment. Ever since the first cave man or woman discovered fire, humans have been in a lifelong struggle to better adapt to their environment. We have forever been in the business of trying to influence biological and psychological nature and to fashion them into beneficial directions and outcomes.[66] Moreover, this altering of nature to meet perceived needs has produced both good and bad results. As medicine progressed, patients and their treatments improved. However, Francis Bacon, the sixteenth-century philosopher, warned readers that they may well die of the cure as of the disease, indicating that at least on some important level medicine in its infancy was more bust than boon. In every way, however, medicine has necessarily insinuated itself into our lives and improved with each passing century, though not without victims who often outnumbered its successes.

Technology, however, remains a different matter, and how different a matter surely delineates how far we will push, paraphrasing the great astronomer Johannes Kepler, "to think God's thoughts for him" rather than after him. In the presence of a surely fatal disease, somatic gene insertion will, given the dire circumstances, potentially do more good than harm. In the case of enhancement, however, it is very unclear whether such insertion will do more good than harm or only harm alone.[67]

Clearly, there seems to be a place for genetic engineering in the curing of diseases or the alleviation of pain associated with them. About this almost no one disagrees. If patients can be cured of cancer, freed from diabetes, liberated from the threat of sickle cell anemia, or live lives free of Alzheimer's or even its threat, the voice of most inhabitants in the world would be a loud clamor of assent. But well is seldom ever left alone, as one wag put it, and so we press for more. We not only want to be free of disease; we would like never to be victims of it. We want the possibility of disease removed. In order to arrive at that utopia, we must play God by devising ways of acquiring genes with an "any means necessary" mindset. The Devil, we discover, really is in the details. Getting from point A to point B requires making life and death decisions for present and future generations and, for the latter, before they have been born.

Playing God reminds us that science is a moving target. Seventy-five years ago, when a person was stricken with, say, the fatal Lou Gehrig's disease (amyotrophic lateral sclerosis), no moral or ethical concerns were raised because only one treatment existed and that one was quickly fatal.[68] Those unfortunate individuals could be spared some pain (though certainly not all) but died within two years of their diagnosis. Today, however, individuals stricken with Lou Gehrig's disease still face ultimate death but generally survive far more than two years (barring a life-threatening complication from secondary infections) by relying on medicines and machines that keep them alive.

The tragic case of Terri Schiavo raises the same moral and ethical complications. Ms. Schiavo suffered from severe bulimia that caused a severe potassium deficiency. A heart attack followed, with several complications that left her in what the courts described as a "persistent vegetative state."[69] Fifty years ago, people like Terri Schiavo did not become a part of common discourse because there would not have been any technology available to create the occasion for an extended conversation. However, because we now have the knowledge (but not necessarily the ethical framework) with which to address such consequences, we find ourselves in a moral quagmire.

The quagmire creates a new and baffling conundrum: what does it mean to be human?[70] This is not necessarily a bad thing as long as we don't define ourselves (or our mothers, fathers, sisters, or brothers) out of the question. It also inadvertently raises questions about the differently abled. If we define humanity as whatever we can make it, then it changes how we view what we have customarily called normal. While on the surface this may or may not be a bad thing, it does pose new issues that do not have easy resolution. Arriving at a reasonable conclusion may require us to enter arenas that are extremely uncomfortable for us and may not bode well for the future of our species.

What we face are not necessarily evil deeds done by evil people, but acts with unknown outcomes practiced by well-meaning, well-intentioned people.[71] The real difficulty comes when unknown outcomes turn into nightmarish realities. Determining faults and affixing blame (in addition to, without doubt, settling monetary claims from class action suits) could well lead to our national undoing.

Before too long, issues we will have to resolve regarding playing God will focus on when genetic interventions should occur, why we feel they are necessary, and devising some means for second-guessing our best intentions and assessing both risks and benefits.[72] Weighing risks and benefits will be especially important and will involve not only assessing the risks or benefits to the recipients of genetic intervention but also determining what risks and benefits will accrue to future generations.

Just how far should scientists be allowed to explore the secrets of life and play God?[73] In all cases? In some cases? Even setting aside the more controversial issues of cloning and stem cell research, the matter does not disappear. Recall in Chapter 4 that we pointed out that as many as 70 percent of all mainstream grocery stores carry unlabeled (at least in the United States) genetically modified foods. What long-term impact will these foods have, if any? Some argue there will be no difference except better nutrition. Others contend that we will certainly see some changes in health (necessarily for the better) in the future.[74] It does not help much that new evidence tells us that cloned meat and cloned milk really are almost the same as real meat and milk.[75] If so, why bother? Again, even setting this matter aside, there remain genetic engineering enhancement issues with respect to plants (especially flowers) and seeds used to farm in arable-unfriendly lands in the Third World.

If all of this were not enough, the issue of patenting life remains. Since it has already been resolved in *Diamond v. Chakrabarty*, the issue of its legality is no longer under discussion. What remains under debate, however, is what happens when an especially successful and highly lucrative genetic intervention is created and its owner allows its use only to the highest bidder or only to certain groups. As Krimsky points out, "[Guiding] biotechnology safely through a future path of potential liabilities represents an important moral responsibility for the public sector.... [G]ood intentions are not sufficient in themselves."[76] Some have argued that if Dr. Frankenstein were alive today, not only would he not have been routed from town but would have been hailed as a hero while he patented his monster.[77] We do not want to run the risk of making human body parts another market commodity on the NYSE.

Lastly, there remain issues of privacy regarding genetic data. Once genetic databanks are in place, this data can be viewed by, of course, the individual; but what measures are in place now to limit this viewing to only those whose genetic data it is? Furthermore, if genetic testing becomes commonplace, is there a moral obligation for an individual to know what potential future his or her genes may hold?[78] That is, if a person is tested and his or her genes reveal underlying serious illnesses, should that person be informed or be prevented from having children? Moreover, will insurance companies demand access to this data, basing their arguments on the right to cover individuals whose health will lead to untoward medical costs?[79] Untying this Gordian knot will prove especially perplexing while remaining critical to the success or failure of public acceptance of genetic engineering. Refusing to untie (or at least attempting to unravel) most of these issues will place us on a road that could lead us to inevitable destruction.

While playing God is, of course, a metaphorical phrase, the promises of genetic engineering appear to want to draw an exact parallel. Genetic engineering promises more and better farming, especially in hard-to-farm-areas; it promises what could be called designer food by making the food favorites of the Western world less likely to contribute to obesity; it promises a better environment and fewer hazardous chemicals and pesticides; genetic engineering promises petroleum-eating bacteria and even better computer technology; it promises supervaccines that will rid the world of all disease; it promises designer babies; and it promises to replace old bodily organs with new ones, either newly created or refashioned from old ones. Finally, genetic engineering promises to make the world a better, safer, and happier place.[80] Given all these promises, it's hard not to believe that genetic engineering will not play God; indeed, it may well want merely to play him to supplant him altogether.

QUO VADIS?

By now, many readers doubtless feel like that poor Cambridge city council member in the 1970s referred to in the Preface to this book. After he heard from both sides on the genetic engineering question (each side bringing forward distinguished scientists, each with conflicting data but all with very convincing graphs, charts, and supportive research), he cried, "What the hell am I supposed to think?"[81] It is a familiar *cri de coeur* and anyone who feels that way is fully justified. Some recommendations may be put forward, however, that may provide useful insight and direction in thinking about the future of genetic engineering. These recommendations will not only make the research safer but will also provide tighter controls on what is being done and by whom.

1.   *All labs doing recombinant DNA research should be reviewed biannually.* Each lab should be physically inspected to be certain waste is being properly disposed of and in a secured fashion. Any lab found not up to par should be shut down immediately and its funding rescinded.
2.   *All research into diseases caused by single genes should continue unabated.* Since most orphan diseases are caused by a single gene, and since single-gene diseases have the greatest promise of success, this should continue, and with additional federal funding.[82]
3.   *All human cloning research should be banned.* We have no idea what we are doing in this area, we have limited chances for success, and there is enormous opportunity for evil even if success be found. Human chimeras, clones for spare human body parts, or clones for research (even if successful) would create a moral and ethical crisis not unlike the research done by Nazi physicians

during World War II. No good appears to be present in this kind of so-called Frankensearch.

4. *Adult stem cell research should receive a greater share of genetic engineering funding to ascertain its potential for success.* New programs that show great promise, such as the work being done by Treena Arinzeh, should be hotly pursued.[83]

5. *All genetically modified foods should be labeled as such.* This is not a very complicated process and should be done immediately. If the American public doesn't want to buy genetically modified foods, they should be given the right amount of information to make that choice.

6. *More research must be done on the long-term effects of genetically modified foods.* While current evidence indicates that genetically modified foods are safe, we still do not know the effects of prolonged consumption. For years scientists have debated the merits and deficits of a seemingly innocuous morning pick-me-up, coffee. It is apparently simultaneously deadly, healthy, a prophylactic against cancer, and a cause of heart disease. Surely more research about genetically modified foods is in order.

7. *More research on transgenic animals is required. Special attention should be paid to the ethical dimensions of pharming.* Animal testing is a staple of medical research and is in no way disputed here. It makes far more sense to test various outcomes on animals than on human populations but creating chimerical animal species just because we can, or creating mice with human ears just to see if it can be done, appears to step into a realm that requires more research on its ethical dimensions before proceeding unchecked.

8. *Cease any biochemical research for purposes of war.* Biochemical warfare has already been outlawed since 1972, when the United States signed the Biological Weapons Convention. The uses of biochemical weapons now via methodologies engendered by genetic engineering are no different than using mustard gas or any other chemical to attack combatants. Even such highly genetically developed chemicals that would allow the extermination of targeted populations appear to be in violation of the Biological Weapons Convention.[84] All such research should be banned, or at least halted, until we can better understand its long-term effects. We thought, for example, that we fully understood the uses of both atomic and hydrogen bombs at the conclusion of World War II. Though such bombs ended the war with Japan and certainly saved ten of thousands (if not millions) of American soldiers' lives, we did not fully comprehend the long-term damage of these deadly weapons. Once we did, we took steps to eradicate their uses altogether and developed far more effective smart bombs that target enemy sites while minimizing human collateral damage. Such research should be encouraged while biochemical warfare should be halted, if not banned entirely.

9. *All enhancement and elective uses of genetic engineering should be banned.* Enhancement genetic technologies serve only small, privileged populations while changing what we consider both human and natural. Enhancement procedures to improve either mental or physical abilities, or both, serve no one well and threaten future populations. It minimizes our human dignity and

makes us the tools of our tools. Enhancement can be achieved well enough through old-fashioned means such as exercise, diet, and study. Quick surgery can just as easily enhance a given ability as destroy it, both in the individual on whom it is performed and in all his or her future progeny. If enhancement technologies are finally perfected in some manner, they should be made freely available and not just to the privileged.

10. *Establish a private or national body to oversee the ethical boundaries of stem cell research; this entity should be invested with the power and authority to halt or ban specified research that exceeds the boundaries of human dignity.*[85] Some agency or organized body must be vested with the right and authority to oversee the ethical boundaries of all stem cell and recombinant DNA research, and have the power to halt or ban any research that exceeds these boundaries.[86] Human cloning, the creation of transgenic animals for pharming beyond what medical science requires, research that is cruel and unusual in its disregard for animal pain, and the like, would be oversight targets for this body. Without such a body, research will continue unabated, forcing us to create the ethical and/or moral response after the fact rather than before.

Genetic engineering has opened Pandora's box. It is the Icarus to science's Daedalus. It remains to us to see to it that, with the lid now lifted, hope remains, and with our wings firmly in place, we fly in formation at the behest of our creator, not into the blazing sun to meet our certain and unavoidable doom.

Appendix: Opposing Views

A topic as complex as genetic engineering does not have a pro/con side the way most controversial topics do. Rather, with genetic engineering it's more nuanced. Views differ more to the extent to which one should go in this uncharted territory, not whether there should be a pursuit at all. The two correspondents responding here are representative of this. While both agree that genetic engineering is potentially a good thing, they disagree about the extent to which is should be vigorously pursued.

Peter Singer is no stranger to controversy, or to the subject of genetic engineering. Mr. Singer served as the Ira W. DeCamp Professor of Bioethics, University Center for Human Values at Princeton from 1994-2004. He assumed a part-time role in that capacity in 2005, sharing the other half of his time as the Laureate Professor at the University of Melbourne in the Centre for Applied Philosophy and Public Ethics. He has published more than two dozen books, many in multiple translations, and scores of articles. His newest book is *In Defense of Animals: The Second Wave.* (Oxford, 2005). Mr. Singer has also appeared on numerous television and radio programs. His well-known view that places animal life at, or equal to, some human life is both controversial and provocative. His contribution here, "Genetics Engineering: An Overview" argues for regulation and some minimal restraints.

Who better to counter a Princetonian than a fellow Princetonian? Robert P. George is the McCormick Professor of Jurisprudence and Director of James Madison Program in American Ideals and Institutions at Princeton. Mr. George is a constitutional law and jurisprudence expert and has written widely, not only on legal matters but also on ethics, genetic engineering, and cultural issues. He holds the 2005 Bradley Prize for Intellectual and Civic Achievement. His books include *In Defense of Natural Law*, *Making Men Moral: Civil Liberties and Public Morality*, and *The Clash of Orthodoxies: Law, Religion and Morality in Crisis*. He has written scores of articles and has been on television and radio numerous times. In addition to all these things, he also serves as a member of the President's Council on Bioethics (see http://www.bioethics.gov/). Mr. George's contribution here, "Genetic Engineering: Risks and Alarums" is a perfect counterpoint to Mr. Singer's overview.

I take this time to thank both men for their studied contributions in this volume.

"GENETICS ENGINEERING: AN OVERVIEW"
PETER SINGER

Genetic engineering is a broad term that embraces a host of different activities. The debate about genetic engineering tends to be highly polarized and polemics often take the place of argument. That is a pity, because there is no more important and ethically challenging set of questions facing us today.

So far, the most significant use of genetic engineering has been to create genetically modified plants for commercial production. In 1996 there were only 4 million acres planted with GM crops worldwide. Seven years later that had risen to 167 million acres. In the United States today, unless absolutely everything you put in your mouth is certified organic it is very difficult to avoid eating some GM products.

The most fundamental ethical objection to GM crops is that genetic modification is a form of human arrogance, almost like playing God. It is wrong, some say, for one species to tamper with the nature of the different species by, for example, inserting a gene from a fish into a plant so as to create an entirely new kind of plant. The second major argument is that GM crops pose an unacceptable risk of irreversible environmental damage.

The first of these arguments can take either a religious or a nonreligious form. If life and all the species are seen as God-given, then for humans to set about to modify them may be regarded as a blasphemous attempt to improve upon God's creation. Without invoking religious beliefs, a similar argument can be put in terms of the intrinsic value of nature and the belief that we should not alter it. Nature itself, rather than God, then becomes sacrosanct. However, it isn't easy to see why both the religious and the nonreligious forms of the argument should not also rule out the kind of selective breeding that has, over many thousands of generations, transformed wild animals into the familiar domestic animals we have today. God created the Burmese jungle fowl, then we transformed it into the modern White Leghorn chicken. Is that blasphemous? If GM corn is unnatural then so, too, is a turkey with a breast so large that it cannot mate and can only reproduce by artificial insemination with human assistance. Why should one way of changing species be considered playing God or contrary to nature, and the other not? Does the fact that it took many generations to breed such turkeys make it more natural and therefore acceptable? Why should the passage of time make a difference? The nineteenth-century philosopher John Stuart Mill thought that appeals to "nature" were frequently a source of "false taste, false philosophy, false morality, and even bad law."[1] That generalization holds in this instance, at least. Unless we are to turn our backs to the domestication of plants and animals and revert to being hunter-gatherers, we cannot seriously think that interfering with the nature of species is intrinsically wrong.

Although interfering with the nature of species is not intrinsically wrong, our awareness of how much we still have to learn about natural processes (whether it is a question of genetics, our own health, or our planet's ecology) might properly make us think that making rapid, novel changes in the nature of species is a risky and unwise thing to do. One of the risks is to the environment. Of course, there may be nothing wrong with eating GM foods but there are certainly some environmental risks from the genes of GM plants crossing with wild relatives. Repeated revelations about regulatory failures with GM crops don't give much confidence that adequate safeguards are in place to prevent this from happening. There is, therefore, an ethical case for avoiding GM because we should not support growing crops or releasing animals where there is even a very slight risk that this could cause an environmental disaster. To take this view is not to say that it is, in principle, wrong to modify any organism or even to say that there will not, in the future, be GM organisms that can be proven to be safe and will bring sufficient benefits to justify commercial use. An effective regulatory system should consider each case on its merits. Since the developed countries can produce an abundance of food without using GM techniques, however, we do not need to take big risks to produce more food. The balance of costs and benefits could be different for developing countries that have a greater need to produce more food.

If we turn to genetic engineering in human beings, the same broad, ethical approach can be used. That is, we should not see the present genetic endowment of the human species as sacrosanct or inviolable. That genetic endowment is the product of millions of years of evolution. It may seem arrogant to believe we can improve upon it but we need to remember that evolution is an unplanned process; it is not, in itself, either good or bad. So if evolution has resulted in, for example, humans who carry genetic diseases, and we could safely eliminate these diseases from the gene pool (without their removal having other negative consequences), it would be good to do so. One difficult question is whether we really could be sufficiently confident that the removal of a particular gene would have no negative consequences.

More controversial and more philosophically significant is the question of whether it could ever be right to enhance human nature beyond what is normal. For example, instead of merely removing genes that cause disease or premature death we could enhance the intelligence of future generations or the extent to which they are likely to act altruistically. Again, we should not believe that there is anything intrinsically wrong with seeking to enhance the nature of human beings. There is nothing sacred about the way we are at this particular moment in our evolutionary history. The important issues concern the consequences of our actions. Many people fear state control over the genetic endowment of future generations. That danger could be eliminated by leaving such choices up to parents. However, the idea of parents making unregulated choices for their offspring conjures up all kinds of unfortunate images, such as a generation of children who resemble pop stars or sports heroes.

Hence, some regulation seems desirable. We need to think about what kind of regulation we want. As we gain a better understanding of the genes that have an impact on qualities such as intelligence and sporting ability (which, of course, are also strongly influenced by our environment), it seems likely that clinics will begin to offer genetic testing of preimplantation embryos for those who can afford them. Some will object to this because the unwanted embryos will be disposed of. That is not my objection; I do not think that an early embryo is the kind of entity that has a right to life or a claim to be protected. For that, at a minimum, some degree of consciousness is required.[2] But there is something objectionable about a society that allows rich parents to genetically select advantages for their children while the poor are unable to do the same. If we value equality of opportunity, we should not allow that to happen. Yet it seems strange to allow rich parents to send their children to expensive preparatory schools and at the same time to prohibit them from using more effective and less costly methods of enhancing their children's scholastic abilities. There is a strong case for saying that access to such forms of genetic selection should either be available to all or to none.

If human genetic enhancement—whether by genetic selection or by genetic modification—becomes feasible and some countries take it up successfully, it will be difficult for anyone to reject it. No country would want to become a scientific and technological backwater because it had refused to enhance the abilities of its future generations. We may have to make our way forward into this ethical thicket even though we cannot know exactly where we will emerge. But that is not grounds for despair, just for the most careful forethought to maximize our chances of ending up in a better place than where we were when we entered the thicket.

"GENETIC ENGINEERING: RISKS AND ALARUMS"
ROBERT P. GEORGE

The day may come when biotechnology makes it possible for parents to custom design their offspring, manipulating genes to produce children with the "superior" traits—strength, intelligence, beauty, etc.—the parent or parents desire. But that day is still a long way off. The relationship between genes and qualities such as intelligence and athletic prowess turns out to be so complex that the dream or nightmare of "designer babies" may never become a reality. That doesn't mean we shouldn't worry about the possibility. But, for now, we shouldn't spend too much of our worry budget on it. There are far more urgent things to be concerned about today in the field of biotechnology.

Before discussing these things, however, we should pause to reflect on the blessings that genetic knowledge and the biotechnologies it makes possible have delivered or will deliver soon. Much genetic knowledge has been generated by inquiry aimed at curing diseases, healing afflictions, and ameliorating suffering. Valuable biotechnologies have been developed for the purpose of advancing human health and well-being. This is to be applauded.

Moreover, genetic knowledge, like knowledge in other fields of intellectual inquiry, is intrinsically valuable. Even apart from its utility in medicine, such knowledge is humanly fulfilling and, indeed, fulfilling in a special way since much genetic knowledge is a species of self-knowledge. Advances in genetics help us to explore and understand more fully that greatest of mysteries, namely, the mystery of man himself. These advances, too, deserve our applause.

Now let us turn to the worries—the urgent ones.

The first worry is that we may compromise, or further compromise, the principle that every human being, irrespective of age, size, mental or physical condition, stage of development, or condition of dependency, possesses inherent worth and dignity and a right to life. Proponents of research involving the destruction of human beings in the embryonic stage for biomedical research began by proposing only that "spare" embryos being held in cryopreservation in IVF clinics be sacrificed. These microscopic humans would, they argued, likely die anyway, so nothing would be lost (and no wrong would be done) by destroying them to harvest stem cells. Soon, however, many of these people were calling for the mass production by cloning of human embryos precisely for use as disposable research material. For now, most insist that they desire to use only embryos in the blastocyst (5-6 day) stage, and are not proposing to implant and gestate embryos that would then be killed at later stages of development to harvest cells, tissues, or organ primordia. But this is bound to change. Having abandoned the moral norm against deliberately taking innocent human life, many will be carried by the logic of their position to the view that producing human beings to be killed in the fetal and even early infant stages is justified in the cause of regenerative medicine.

The second worry is closely related. It is that many people are coming to view procreation as akin to manufacture and to regard children, not as gifts to be cherished and loved even when "imperfect," but rather as products that may legitimately be subjected to standards of quality control and discarded or killed in the embryonic, fetal, and even infant stages if they do not measure up. Preimplantation genetic diagnosis (PGD) of embryos in the context of assisted reproduction is increasingly widely practiced. In IVF clinics in the United States, it is common for a larger number of embryos to be produced than can be safely implanted. So, people reason, why not choose the ones likely to be healthiest? Embryonic human beings are considered more and less worthy of life, and sometimes not worthy of life at all, depending on their "quality." And the eugenic ethic embodied in the practice of PGD is not confined to choosing among embryos for implantation. Eugenic abortion—and, in some cases, even infanticide—is regarded as perfectly legitimate by many in the United States and elsewhere. A child in the womb who has been diagnosed with Downs' syndrome or dwarfism is likely to be aborted. A newborn may be deprived of a simple life-saving surgery and "allowed to die." Those responsible will, perhaps, tell themselves that they are doing it "for the good of the child." The reality, however, is that they are treating the retarded or handicapped child as a "life unworthy of life." And let no one suppose that such decisions, ghastly as they are even when chosen by parents, are or will remain a matter of unencumbered "choice." Social pressures exist and will build for parents to spare society the burdens of caring for, or even encountering, handicapped or retarded people. Some years ago, the geneticist Bentley Glass, envisaging a future in which genetic screening would become the routine thing it is today, proclaimed triumphantly that "no parent will . . . have a right to burden society with a malformed or a mentally incompetent child."

The great bioethicist Leon Kass has diagnosed the situation insightfully. Speaking at the United States Holocaust Museum, Kass warned:

[The] eugenic vision and practice are gaining strength, all the more so because they grow out of sight behind the fig leaf of the doctrine of free choice. We are largely unaware that we have, as a society, already embraced the eugenic principle 'Defectives shall not be born,' because our practices are decentralized and they operate not by coercion but by private reproductive choice.

One should observe, of course, that many people continue to resist the eugenic ethic and struggle to reverse it; and despite the (sometimes amusing) boasting of the eugenicists, there is no good reason to think that it cannot be reversed in significant measure. Yet a sober assessment of the situation requires us to acknowledge that support for the eugenic killing of human beings in the fetal and infant stages is no longer a "fringe" position, and is particularly strong in elite sectors of the culture.

Groups dedicated to defending the dignity and rights of handicapped persons (even when they take no official position on the ethics of abortion as such) have recognized the dire implications of the eugenic ethic for the persons they serve. As Dr. Kass puts it, "persons who happen still to be born with these conditions, having somehow escaped the spreading net of detection and eugenic abortion, are increasingly regarded as 'mistakes,' as inferior human beings who should not have been born." This has produced an alliance between the pro-life movement and advocates of justice for the handicapped or disabled in a number of domains.

The glory of our political tradition is its affirmation of the profound, inherent, and equal dignity of all human beings. The history of our politics and social practice, our law and economics, and even our medicine is in significant measure the struggle to live up to the demands of this affirmation. The trouble, of course, is that individual and collective self-interest are often at war with it. All-too-often, people will have powerful motives to regard others as less-than-fully-human, or to believe that humanity can be divided into classes—superiors and inferiors, "persons" and subpersonal or non-personal members of the human family. It was true in the days of slavery; it is true in the era of eugenic abortion and infanticide. Sometimes people say that the challenges of biotechnology will require us to invent new principles of ethics and politics. At least when it comes to the immediate dangers we face, that isn't true. What we need is fidelity to the principles of human equality and dignity that have always served us well when we have had the wisdom and fortitude to honor them.

Notes

PREFACE

1. This can be arrived at by various means. I chose to do it as follows: I searched the Online Computer Library Center (OCLC) database for titles published between 1980 and 1995 and then searched it again from 1996–2005 (June). There is a nearly 20% jump in the past 9 years over the previous 15, indicating a fairly dramatic level of increased interest. It also spelled trouble for the researcher who hoped to read everything.

2. Rowan Hooper, "Genes Blamed for Fickle Female Orgasm," *New Scientist,* 8 June 2005 (accessed via www.newscientists.com/article.ns?id=dn7481, June, 2005). Apparently neither size nor anything else matters but genes.

3. Daniel E. Koshland Jr., "Sequences and Consequences of the Human Genome," *Science,* (13 October 1989) 246 (4927): 189. This was a much-talked-about editorial when it appeared. Koshland has backed a way from this now, allowing that personal choice may also factor into social problems.

4. For example, hypertension or high blood pressure. Although this is taken up again in Chapter 7, suffice it to say for now that while we know genes have much to due with hypertension, changing that one gene will do little if we cannot change the eating propensities of those same individuals.

5. See David Chazan, "Who Are the Raelians?" *BBC News,* Saturday, 28 December 2002 (accessed via news.bbc.co.uk/1/hi/health/2610795.stm, June, 2005). I take up this interesting group and their antics again in Chapter 5.

6. Scylla and Charybdis are two sea monsters from Greek mythology. In Homer's story (*The Odyssey*), these two beasts blocked Ulysses's attempt to get back home following his long trek in Troy. While Ulysses and his men escaped the efforts of Charybdis to drown them in a whirlpool, they could not avoid the many-headed beast, Scylla, who

came out of nowhere and took six of his men. See www.2020site.org/ulysses/scylla.html for more (accessed June, 2005).

CHAPTER 1
"It's Alive!" Public Perceptions of Genetic Engineering

1. Brigette Nerlich, David D. Clarke, and Robert Dingwall, "Fictions, Fantasies, and Fears: The Literary Foundations of the Cloning Debate," *The Journal of Literary Semantics* 20, no. 1 (2001): 38. Emphasis added.

2. See David J. Skal's excellent *Screams of Reason: Mad Science and Modern Culture* (New York: W. W. Norton & Company, 1998), 21.

3. Richard Shattuck, *Forbidden Knowledge: From Prometheus to Pornography* (San Diego: Harcourt Brace, 1997). Shattuck's worry is about a different kind of knowledge, in this case the perverted knowledge of De Sade whose work occupies the first half of the book. The matter, however, is very much the same: there are those who seek knowledge solely on the basis that we can, at times, do a thing without regard to what might be the eventual result. For Shattuck, there are things we can do that we should not, and there is a kind of knowledge that we can posses that we should most assuredly run from. In case readers have trouble thinking of examples in this latter context, ponder for a moment the so-called science that Nazi physicians attempted to collate before the end of World War II, in which they tortured Jews for no apparent reason other than to see what pain could be exacted. See Robert Jay Lifton's *The Nazi Doctors: Medical Killing and the Psychology of Genocide* (New York: Basic Books, 1986). Dante gets at this notion (more at knowing in general than knowing the forbidden) in his *Purgatorio: A Verse Translation,* trans. Jean & Robert Hollander (New York: Doubleday, 2003), 60, note 37. Hollander's note is illustrative here but grates on modern sensibilities. *Benvenuto da Imola* paraphrases Dante: *Sufficiat vobis credere quia sic est, et non quaerere propter quid est.* "Let it suffice you to believe that something is so, without seeking to know why it is so." Could it be that our desire to know may blind knowers to potential bad outcomes?

4. While it's true this movie has been remade, the original 1958 version focuses on an invasion by the result of some scientific experiment gone horribly awry.

5. Quoted in David Skal, 34. I have relied on Skal's account but there are many in any number of literary histories.

6. *Isaiah* 14:13–15. The verse runs, "For thou hast said in thine heart, I will ascend into heaven, I will exalt my throne above the stars of God: I will sit upon the mount of the congregation, in the sides of the north." In other words, I will not merely play God; I will be God.

7. Skal, 69.

8. Nerlich, 44.

9. Skal, 68.

10. Robert Louis Stevenson, *The Strange Case of Dr. Jekyll and Mr. Hyde and Other Famous Tales* (New York: Dodd, Mead & Co., 1961), 6. Also quoted in Skal, 69.

11. Skal, 103.

12. Ibid.

13. Ibid., 104.

14. See Peter Dans's *Doctors in the Movies: Boil the Water and Just Say Aah* (Bloomington, IL: Medi-Ed Press, 2000), 3.

15. Ibid., 35.

16. Ibid., 99, 107.

17. Skal mentions *The Man Who Killed Death, The Head* (1959) and one of my all-time favorites, *The Brain That Wouldn't Die* (1962), in which the very comely but fierce Virginia Leith's head alone (drat) lives in a petri dish for the duration of the movie.

18. Dans, 18–19.

19. Ibid., 195.

20. Quoted in Charles Weiner, "Drawing the Line in Genetic Engineering: Self-Regulation and Public Participation," *Perspectives in Biology and Medicine* 44, no. 2 (Spring 2001):208 (accessed via *InfoTrac, Expanded Academic ASAP Plus*, August, 2004.)

21. Ibid.

22. Zac Goldsmith, "Who Are the Real Terrorists?" *The Ecologist* 28, no. 5 (September-October 1998): 312 (accessed via *InfoTrac, Expanded Academic ASAP Plus*, August, 2004.)

23. Ibid.

24. Ibid.

25. Kurt Bayertz, *GenEthics' Technological Intervention in Human Reproduction as a Philosophical Problem* (New York: Cambridge University Press, 1987), 77 (accessed via *InfoTrac, Expanded Academic ASAP Plus*, August, 2004.)

26. For this view, namely, that Europeans are more skittish than Americans about the prospects of genetic engineering, see Isaac Rabino, "The Biotech Future." *American Scientist* 86, no. 2 (March–April 1998): 110–113 (accessed via *InfoTrac, Expanded Academic ASAP Plus*, August, 2004).

27. Ibid., 110.

28. *Wellcome Trust Report* (1998) "Public Perspectives on Human Cloning," *Medicine in Society Programme,* online PDF at ag:www.wellcome.ac.uk. 6.2 (accessed August, 2004). Also quoted in Nerlich, 44.

29. Alan L. Otten, "In Poll About Genes, Most Say That Ends Often Justify Means," *Wall Street Journal* (May 28, 1987): 34 col. 2.

30. 2003 Roper Center at the University of Connecticut Public Opinion Online. The question was, "I'm going to describe a few of these developments (in science and medicine) that have been in the news and would like you to tell me how much you have heard or read about each of them ... Genetic engineering, a technique to change DNA or the building block of life, in order to produce particular characteristics ... How much have you heard or read about this—a great deal, something but not very much, or nothing at all?" (accessed via *LexisNexis Academic*, October, 2004).

31. Ibid. All of the polling data in the next paragraphs are from this source unless otherwise noted. For clarity, questions are noted here if pertinent.

32. This question comes from a 1986 Louis Harris and Associates Poll but is found at the same source listed above.

33. While we cannot know and it would be wrong to speculate, it would be interesting to know if any of this 11 percent are scientists, or work in the area of biotechnology, or stand to gain by unhindered biotechnology.

34. This question and the next three are found in the same online source listed in Note 30, but come from a question asked by Princeton Survey Associates in 2002.

35. We will address this again in Chapter 4. For now, superweeds are weeds that, owing to crosspollination with genetically modified plants, grow in the wild and are impossible, or nearly impossible, to kill. This question and the ones below are from a nationwide Zogby International Poll of American adults in January, 2002. The information is found at the same online sources as listed in Note 30.

36. Louis Harris Poll, October, 1986, found in the same online source listed in Note 30.

37. See, for example, Keith Schneider, "Public of 2 Minds on Genetic Shifts," *New York Times,* May 31, 1987 (Late City Final Edition), sec. 1, part 1, page 21, col. 1.

38. Alexander Gorke and Georg Ruhrmann, "Public Communication between Facts and Fictions: On the Construction of Genetic Risk," *Public Understanding of Science* 12, no. 3 (2003): 229.

CHAPTER 2

History of Genetic Engineering from Mendel to (Genome) Maps

1. See C. S. Lewis, *Allegory of Love: A Study in Medieval Literature* (New York: Oxford University Press, 1958). *Sic passim,* but especially 3–43. Also, *An Experiment in Criticism* (Cambridge: Cambridge University Press, 1961), 43–44 specifically, but all of Chapter 5, "On Myth," is pertinent here.

2. I have taken the telling of this myth from *The Metamorphoses of Ovid,* trans. Mary M. Innes. Harmondsworth, Middlesex, England: Penguin Books Inc., 1955, 184–185.

3. The emphasis is my own, though the words are from Ovid.

4. The original story has it as a vase, not a box. However, owing to the Dutch author Erasmus's use of a box in 1508, it has been a box ever since.

5. This is a rare genetic disorder caused by a deficiency of an enzyme (hypoxanthine-guanine-phosphoribosyltransferase), mercifully HPRT for short. Shortly after birth (3–6 months) Lesch-Nyhan Syndrome is characterized by self-mutilation (lip and finger biting, head-banging). The high levels of uric acid caused by this defect later attack joints, heart, kidneys, and more, often with compulsive behaviors. Moderate mental retardation occurs. There is no treatment and death

generally occurs before the onset of puberty. For more, see www.ninds.nih.gov/disorders/lesch_nyhan/lesch_nyhan.htm (accessed December, 2004).

6. Our use of the word is attributed to William Bateson (1861–1926), dubbed "the apostle of Mendelism in England." See Peter R. Wheale and Ruth M. McNally, *Genetic Engineering: Catastrophe or Utopia?* (New York: St. Martin's Press, 1988), 3–4.

7. J. Levine and D. Suzuki, *The Secret of Life: Redesigning the Living World* (Boston: WGBH, 1993), p. 13. Interestingly, it is the eighth day on which Abraham is told to circumcise. It is also on the eighth day that enough vitamin K is built up in the body that might prevent an injury from leading to certain death of all males. It would appear that however rudimentary their knowledge, the Hebrews still knew quite an astonishing amount for a nomadic people.

8. See Genesis 30:31–35.

9. Aristotle treats this in Book II, *Nichomachean Ethics* and also *Historia Animalium, sic passim.* See David Suzuki and Peter Knudston, *GenEthics: The Clash Between the New Genetics and Human Values* (Cambridge, MA: Harvard University Press, 1989), 45, and www.mlahamas.de/Greeks/AristotleBiol.htm for more (accessed August, 2004).

10. William Shakespeare, *Titus Andronicus,* Act V, Scene 1, lines 27–32. The last two are also quoted in British Medical Association, *Our Genetic Future: The Science and Ethics of Genetic Technology* (Oxford: Oxford University Press, 1992), 6.

11. D. S. Halacy Jr., *Genetic Revolution: Shaping Life for Tomorrow* (New York: Harper & Row, 1974), 9.

12. This and the next word are from the *Oxford English Dictionary.*

13. Nathaniel C. Comfort, "Are Genes Real?" *Natural History* 110, no. 5, June, 2001, (accessed via Thomson Gale's Infotrac, December, 2004).

14. Unless otherwise noted, biographical data about Mendel comes from V. Kruta and V. Orel, "Johann Gregor Mendel," in *Dictionary of Scientific Biography,* ed. Charles Coulston Gillispie, vol. 9, A. T. Macrobius—K. F. Naumann (New York: Charles Scribner's Sons, 1974), 277–283. See also Colin Tudge, *The Engineer in the Garden: Genes and Genetics: From the Idea of Heredity to the Creation of Life* (New York: Hill and Wang, 1993), 10–55.

15. Leeuwenhoek (pronounced lay-wen-hook) invented as many as 500 microscopes, discovered bacteria, viewed animal sperm for the first time, and really made visible the world of life beyond our mere eyes. See, for example, an encyclopedia entry or www.umcp.berkeley.edu/history/leeuwenhoek.html (accessed December, 2004).

16. Mae-Wan Ho, *Genetic Engineering: Dream or Nightmare?* (New York: Continuum, 2000), 75.

17. Marc Lappé, *Genetic Politics: The Limits of Biological Control* (New York: Simon and Schuster, 1979), 12–13.

18. Keekok Lee, *Philosophy and Revolutions in Genetics: Deep Science and Deep Technology* (New York: Macmillan, 2003), 2.

19. William Bains, *Genetic Engineering for Almost Everybody* (New York: Penguin Books, 1987), 10–11.

20. The table is a derivation from Karl A. Drlica, *Double-Edged Sword: The Promises and Risks of the Genetic Revolution* (Reading, MA: Addison-Wesley Publishing Company, Helix Books, 1994), 5. Most introductory texts to genetics will have a similar representation. See also William H. Sofer, *Introduction to Genetic Engineering* (Boston: Butterworth-Heinemann, 1991), chapters 3 and 4.

21. Suzuki and Knudston, *GenEthics,* 34.

22. Ho, *Genetic Engineering,* 72.

23. R. Cole-Turner, *The New Genesis: Theology and the Genetic Revolution* (Louisville, KY: Westminster/John Knox Press, 1993), 13.

24. Edward I. Alcamo, *DNA Technology: The Awesome Skill* (New York: Harcourt Press, 2001), 6.

25. Ibid., 73. See also British Medical Association citation in Note 10, 12–13.

26. Halacy, 13, raises this point.

27. David Fairbanks and Bruce Rytting, "Mendelian Controversies: A Botanical and Historical Review." *American Journal of Botany* 88, no. 5 (May, 2001): 737–752, especially pages 738, 743–745.

28. P. R. Reilly, *Abraham Lincoln's DNA and Other Adventures in Genetics* (New York: Cold Spring Harbor Laboratory Press, 2000), 16.

29. J. Lyon and P. Gorner, *Altered Fates: Gene Therapy and the Retooling of Human Life* (New York: W. W. Norton & Co., 1995), 39.

30. Ibid., 82–99.

31. Quoted in Jane Goodfield, *Playing God* (New York: Random House, 1977), 12.

32. Factual information, which follows, is from Christopher Wills, *The Wisdom of the Genes: New Pathways in Evolution* (New York: Basic Books, Inc., 1989), 22–23. See also Elizabeth Marshall, *High-Tech Harvest: A Look at Genetically Engineered Foods* (New York: Franklin Watts, 1999), 26–45, for a very compelling, easy-to-understand discussion of this very technical process.

33. The figure is mentioned here parenthetically but is important because it is far below the 100,000 to 150,000 originally predicted by the Human Genome Project, about which see Chapter 6.

34. Comfort, 30. The work was done by French geneticists François Jacob and Jacques Monod.

35. Alcamo, 9.

36. Wheale and McNally, 4.

37. Alcamo, 9.

38. Quoted in Lee, 124.

39. Levine and Suzuki, 16.

40. British Medical Association, *Our Genetic Future: The Science and Ethics of Genetic Technology* (Oxford: Oxford University Press, 1992), 30.

41. Levine and Suzuki, 16.

42. See www.time.com/time/time100/scientist/profile/watsoncrick.html (accessed December 2004).

43. F. Crick and J. Watson, "A Structure for Deoxyribose Acid (D.N.A.)." *Nature* 25 April 1953. Available at www.dna50.org/main.htm (accessed December, 2004).

44. Levin and Suzuki, 16–18. What follows relies heavily on these pages unless otherwise indicated. Of course, DNA has a good deal of help, too. See Finn Bowring, *Science, Seeds and Cyborgs: Biotechnology and the Appropriation of Life* (New York: Verso, 2003), 32–34; G.J.V. Nossal and Ross L. Coppel, *Reshaping Life: Key Issues in Genetic Engineering* (New York: Cambridge University Press, 2002), 18–30; and Sandy Bornstein and Jerry Bornstein, *New Frontiers in Genetics* (New York: Julian Messner, 1984), 42–73.

45. French Anderson, "Genetic Therapy." In *The New Genetics and the Future of Man,* ed. Michael P. Hamilton. Grand Rapids, MI: William B. Eerdmans, 1972, 111.

46. See J. Watson, *The Double Helix: A Personal Account of the Discovery of the Structure of DNA* (New York: Atheneum, 1968). Indeed, so brazen was the Nobel Prize winner's minimization of Franklin's role, even after admitting his ignorance of the x-ray crystallography that set him on the right track, that Harvard University Press refused his manuscript. The accepting publisher required an epilogue that essentially said he had understated Franklin's role and that she was, in fact, far more pivotal. See *The Double Helix* and also www.brown.edu/Courses?BI0020_Miller/dh/guide.html (accessed December, 2004). Nobel Prize winner Linus Pauling's work with alpha-helix for proteins was also quite helpful.

47. That the function of genes is to direct the production of enzymes and other proteins was verified in 1941 (a little more than a decade before Watson and Crick's famous paper) by Stanford geneticist G. W. Beadle and biochemist E. L. Tatum, both Nobel Prize winners, 1958. See Lyon and Gorner, 41.

48. The scientists Michael Behe and William Dembski come to mind.

49. Watson and Crick, 3.

50. This example comes from Lee M. Silver, *Remaking Eden: How Genetic Engineering and Cloning Will Transform the American Family* (New York: Avon Books, 1998), 30. See also Michael J. Reiss and Roger Straughan, *Improving Nature? The Science and Ethics of Genetic Engineering* (New York: Cambridge University Press, 1996), 14–15.

51. Stefan-M. Puslt, *Neurogenetics* (New York: Oxford University Press, 2000), vii. These can come from either the mother or father, of course.

52. J. Etienne-Decant, *Genetic Biochemistry: From Gene to Protein* (New York: John Wiley & Sons, 1998), 146.

53. This portion of the discussion relies, among other sources, most heavily on P. Schimmel and H. E. Wendler, "Genetic Code," in *McGraw-Hill Encyclopedia of Science & Technology,* 9th ed., vol. 7, New York: McGraw-Hill, 2002, 787–790. See also Richard J. Reece, *Analysis of Genes and Genomes* (West Sussex, England: John Wiley & Sons, 2004), 1–10.

54. Helen Kreuzer and Adrianne Massey, *Recombinant DNA and Biotechnology: A Guide for Students* (Washington: ASM Press, 2001), 79.

55. The example is from Dean Hamer, *The God Gene: How Faith Is Hardwired into Our Genes* (New York: Doubleday, 2004), 57.

56. Ibid., 80.

57. Alcamo, 109.

58. See Bains, 68–69, for another similar example.

59. Michael R. Lentz, "DNA Replication." In *Encyclopedia of Genetics: Aggression–Heredity–Environment,* ed. Jeffrey A. Knight. Vol. 1. Pasadena, CA: Salem Press, 1999, 161.

60. Alcamo, 54–63.

61. I have closely followed Walker and McKay, 14–22. I have greatly simplified even their simplification for our purposes. Readers seeking more in-depth information can see Alcamo and others cited.

62. For more on this process and the triplets, see British Medical Association, 32–35. The Lamarck reference in the next sentence is found on page 19.

63. Alcamo, 48, and also the next sentence.

64. Jeremy Cherfas, *Man-Made Life* (New York: Pantheon Books, 1989), 48.

65. Kreuzer and Massey, 84.

66. Ibid., 63, for this and the next three facts.

67. Comfort, 31.

68. Lyon and Gorner, 13.

69. Wheale and McNally, 10, for this and the fact in this sentence and the next.

70. British Medical Association, 31. Alcamo also notes these and others in his second chapter.

71. Hamer, 58.

72. Consider, for example, the drug protocol that relied only upon an Internet search. The drug protocol was used and resulted in the death of a young woman who volunteered for the trials. A review of the literature discovered that a search that included both the web and conventional digging in paper indexes would have turned up at least two papers warning that the then-used protocol combination can result in respiratory failure in some individuals, yielding the certain fact that the Internet is no substitute for a library or hard work.

73. Richard Monastersky, "51 Years Later, Influential DNA Manuscript to Be Published." *Chronicle of Higher Education* 50, no. 32 at http://chronicel.com/weekly/v50/i32/32a01801.htm (accessed November, 2004). See also the Watson and Crick paper in *Nature,* cited in note 32 above.

74. For a clear-headed discussion of how this works, see John M. Sedivy and Alexander L. Joyner, *Gene Targeting* (New York: W. H. Freeman and Co., 1992), 17–20. This will be addressed again in the next chapter.

75. To be fair, McClintock did consider these to be genes but chromosomal elements. James Shapiro and Sankhar Adhya discovered that certain bacteria DNA cut themselves out and inserted themselves in two different places, confirming McClintock's work. See Comfort, 30.

76. Barbara Wexler, *Genetics and Genetic Engineering* (New York: Thomson Gale, 2004), 8–9. Alcamo's book, cited above, may be the best historical guide to use to come to understand the very many men and women whose work remains unnoted but still ranks as seminal in this study.

77. The specific examples are from Jon Gordon, *The Science and Ethics of Engineering the Human Germ Line* (New York: Wiley-Liss, 2003), 22.

78. Kreuzer and Massey, 134. There are many places to find similar definitions. I chose this one for no particular reason.

79. See Lee, 6, where he uses the example of a silk rose.

80. See Goldenhagen, *sic passim.*

81. See www.c-fam.org/FAX/Volume_3/fav3n32.html (accessed December, 2004).

82. Wheale and McNally, 41.

83. Rick J. Carlson and Gary Stimeling, *The Terrible Gift: The Brave New World of Genetic Medicine* (New York: Public Affairs, 2002), 12.

84. Alcamo, 225.

85. Quoted in Reiss and Straughan, 61.

86. All of the facts in this and the preceding two paragraphs are from "Double Lives." *U.S. News & World Report* 136, no. 7 (February 23, 2004): 40–41, *InfoTrac Expanded Academic ASAP* (accessed December, 2004).

87. Diane B. Paul, *Controlling Human Heredity* (Atlantic Highlands, NJ: Humanities Press, 1995), 64.

88. Allen Buchanan, et al., *From Chance to Choice: Genetics and Justice* (Cambridge: Cambridge University Press, 2000), 27.

89. Quoted in Kurt Bayertz, *GenEthics: Technological Improvements in Human Reproduction as a Philosophical Problem* (New York: Cambridge University Press, 1987), 33. The ensuing section is most interesting.

90. Quoted in Paul, 65.

91. Ibid., 76. The next sentence is also quoted in Paul, 177.

92. These are the words of Charles Davenport, social Darwinist. Quoted in Buchanan, 44. Uncannily, I heard not long ago a very similar argument made about the hopelessly ill. An infant male born with only a part of his heart was undergoing surgery. A middle-aged man lamented all the money being spent and for what—a postponed but still premature death that will cause insurance rates to skyrocket. It did not seem ironic to him that he had just undergone a serious medical procedure that had saved his life. While outside the scope of this book, preachers who shouted the clarion call to a better, purer race in God's name are fully vindicated in Christine Rosen, *Preaching Eugenics: Religious Leaders and the American Eugenics Movement* (New York: Oxford University Press, 2004).

93. Bayertz, 41.

94. Buchanan, et al., 33, also makes this point.

95. Daniel J. Kevles, *In the Name of Eugenics: Genetics and the Uses of Human Heredity* (Cambridge, MA: Harvard University Press, 1995), 299–301.

96. Quoted in Lilly E. Kay, "Problematizing Basic Research in Molecular Biology." In *Private Science: Biotechnology and the Rise of Molecular Sciences,* ed. Arnold Thackray. Philadelphia: University of Pennsylvania Press, 1998, 22. It did not help matters that John D. Rockefeller was himself anti-Semitic. See Ron Chernow, *Titan: The Life of John D. Rockefeller, Sr.* (New York: Random House, 1998).

97. Paul, 30–31. In fact, it was Galton who coined the term "eugenics," calling it "good in birth" and used it to establish "selective breeding" or breeding on those with "good stock."

98. Galton quoted in Kevles, 14.

99. Ibid., 80–84.

100. Mark Y. Herring, *Pro-Life, Pro-Choice Debate* (New York: Greenwood Press, 2003), 53–56. For a positive spin, see www.plannedparenthood.org/about/thisispp/sanger.html. For a less positive spin see www.spectator.org/dsp_article.asp?art_id=6738 (both accessed January, 2005).

101. Kevles, 47.

102. Ibid., 49, for Juke. For Kalliak, 52.

103. George P. Smith, II, *The New Biology: Law, Ethics and Biotechnology* (New York: Plenum Press, 1989), 90.

104. Paul, 11; Chesterton and Kevles Prussianism, 120. Chesterton collected his criticism in a book of essays called *Eugenics and Other Evils*. He warned that this meant, as Kevles points out, "forcible marriage by the police." Chesterton, while always entertaining, was nearly always right.

105. Paul, 14–15.

106. Jane Maienschein, *Whose View of Life? Embryos, Cloning and Stem Cells* (Cambridge, MA: Harvard University Press, 2003), 104.

107. The story is taken nearly verbatim from Philip R. Reilly, *Abraham Lincoln's DNA and Other Adventures in Genetics* (New York: Cold Spring Harbor Laboratory Press, 2000), 303–304.

108. Maienschein, 105.

109. Quoted in Bayertz, 59.

110. Jeremy Rifkin, *Declaration of a Heretic* (Boston: Routledge & Kegan Paul, 1985), 61–62.

111. This very point is made by Laurence E. Karp, *Genetic Engineering: Threat or Promise?* (Chicago: Nelson-Hall, 1976), 49. This same point is made in a political context by the great social philosopher Nicholas Berdyaev.

112. Gregory Stock, *Redesigning Humans: Choosing Our Genes, Changing Our Future* (New York: First Marnier Books, 2003), 176–201.

113. This is not mere speculation. See Ronald M. Green, "I, Clone" in *Understanding Cloning*, ed. Sandy Fritz. New York: Scientific American, 2002, 92–93.

CHAPTER 3
Splicing, Dicing, and Cloning to Asilomar and Beyond

1. See Leon Kass. "Freedom, Coercion and Asexual Reproduction." In *Freedom, Coercion, and the Life Sciences,* ed. D. Callahan and L. Kass, The Hastings Center, n.p., n.d.

2. Aldous Huxley, *Brave New World* (New York: Bantam Books, 1962), 2–3. Huxley's book appeared in 1939. The great upsurge in research into the structure of and understanding the genetic nature of microorganisms did not occur until the 1940s. If all this were not enough, Huxley invented a new drug for his "slaves" in the *Brave New World* that would make them docile. The drug he invented was called "Soma" (meaning body, of course). The *Physician's Desk Reference* (PRR), the Bible for physicians when prescribing medicine, listed just such a drug at one time. See also

Joseph Fletcher, *The Ethics of Genetic Control: Ending Reproductive Roulette* (Buffalo, New York: Prometheus Books, 1988), 190.

3. C. S. Lewis, *The Abolition of Man* (New York: Collier-Macmillan, 1965), 70–71.

4. V. Sgaramella and A. Bernardi, "DNA Cloning." In *Encyclopedia of Genetics,* eds. Sydney Brenner and Jeffrey H. Miller, Vol. 1, A–D, San Diego: Academic Press, 2002, 544.

5. Burke K. Zimmerman, *Biofuture: Confronting the Genetic Era* (New York: Plenum Press, 1984), 24. See also John Lear, *Recombinant DNA: The Untold Story* (New York: Crown Publishers, 1978), 10–11.

6. While other sources are cited, one should not miss Karl Drlica, *Understanding DNA and Gene Cloning* (New York: John Wiley and Sons, Inc., 1997) or his *Double-Edged Sword* (New York: Addison-Wesley, 1994). The first book covers the gamut of the discussion about the process of gene splicing and cloning, as well as applications of same. The second updates the first and expands the discussion of the pros and cons concerning genetic engineering.

7. Sgaramella and Bernardi, 545.

8. D. Carroll. "Recombinant DNA." In *Encyclopedia of Genetics,* eds. Sydney Brenner and Jeffrey H. Miller, Vol. 3, M–R. San Diego: Academic Press, 2002, 1637.

9. For Alcamo Berg, 90; for Boyer and Cohen, "The Birth of Biotech." *Technology Review* 103, no. 4 (July 2000): 120 (accessed via *Infotrac Expanded Academic ASAP Plus,* September, 2004). Boyer and Cohen were rewarded quite nicely with patents worth more than $250 million before the patent ran out in 1997. Boyer later began with Robert Swanson of Genentech, about which more at the end of this chapter.

10. Carl L. Bankston, III. "Genetic Engineering: Social and Ethical Issues." In *Encyclopedia of Genetics,* ed. Jeffrey A. Knight, Vol. 1. Pasadena, CA: Salem Press), 1999, 262. The idea of restriction enzymes as cutting tools is so firmly set that some illustrations (for laypersons and would-be geneticists) show very tiny pairs of scissors cutting the DNA sequence. See Sandy Primose, Richard Twyman, and Bob Old, *Principles of Gene Manipulation,* 6th ed. (Malden, MA: Blackwell, Science), 28.

11. S. Steven Potter. "Recombinant DNA Methods." In *Recombinant DNA Research in the Human Prospect,* ed. Earl D. Hanson, Washington, DC: American Chemical Society, 1983, 22. Of course, like all analogies, this fails, too, because it implies that what one begins with is not good or not good enough. This may or may not be true, depending on what's being done.

12. Stuart Newman. "The Next Four Years, The Biotech Agenda, The Human Future: What Direction for Liberals and Progressives." December 9, 2004; http:www.genetics-and-society.org/events/20041209_postelections/newman.html (accessed March, 2005); www.lifesciencenetwork.com/news-detail.asp?newsID=460 (accessed March, 2005). Scientists are at pains to contest, as in this citation, that the child-like ear was not really human but was seeded and attached.

13. Sgarmella and Bernardi, 546.

14. Jeremy Cherfas, *Man-Made Life* (New York: Pantheon Books, 1982), 75; Alcamo, 76.

15. Craig S. Laufer. "Cloning Vectors." In *Encyclopedia of Genetics*, ed. Jeffrey A. Knight, Vol. 1: Pasadena, CA: Salem Press, 1999, 118–119; Alcamo, 98.

16. Ellen G. Strauss and James H. Strauss. "Eukaroytic RNA Viruses: A Variant Genetic System." In *Exploring Genetic Mechanisms*, ed. Maxine Singer and Paul Berg. Sausalito, CA: University Science Books, 1997, 121–125.

17. Primrose, Twyman, and Old, 38; John Lear, *Recombinant DNA: The Untold Story* (New York: Crown Publishers, Inc. 1978), 22–24.

18. Zimmerman, 46–47.

19. Also quoted in Mike May, "Mother Nature's Way." In *Understanding Cloning*, ed. Sandy Fritz, New York: Scientific American, 2002, 69.

20. James F. Shepard. "The Regeneration of Potato Plants from Leaf-Cell Protoplasts." In *Understanding Cloning*, ed. Sandy Fritz, New York: Scientific American, 2002, 8.

21. Robert P. Lanza et al. "Cloning Noah's Ark." In *Understanding Cloning*, ed. Sandy Fritz, New York: Scientific American, 2002, 24–25.

22. Zimmerman, 25; Lear, 150.

23. Sheldon Krimsky, *Genetic Alchemy: The Social History of the Recombinant DNA Controversy.* (Cambridge, MA: MIT Press, 1982), 13.

24. Zimmerman, 45.

25. Lear, 70.

26. Ibid., 151ff. I have followed Lear here unless otherwise noted. The story appears in many texts on genetic engineering.

27. Lear, 34. This is sometimes referred to as "Adeno-SV40 nondefectives."

28. Krimsky, 58.

29. Lear, 23.

30. Krimsky, 39.

31. Lear, 24.

32. I have followed Lear's account here very closely.

33. It was doubtless long overdue. See Nicholas Wade, *The Ultimate Experiment: Man-Made Evolution* (New York: Walter and Company, 1977), especially chapter 4, "The Origins of the Moratorium."

34. Lear, 103.

35. Christopher Lampton, *DNA and the Creation of New Life* (New York: ARCO Publishing, 1983), 50–60. This book provides a good over view of the debate, Asilomar, and more.

36. There were actually two Asilomar Conferences, one in 1973 and this one. The 1975 Conference is the one most often referred to because of the subsequent NIH guidelines. That Berg played a pivotal role in the 1973 conference made him a logical pick to oversee the 1975 event. See Krimsky, 58–69. See also, Wade, Chapter 5, "The Conference at Asilomar."

37. The "NIH Guidelines" were the eventual product of Donald S. Fredrickson. Fredrickson found himself in the midst of one of the most controversial times that the once sleepy NIH had ever seen. The "Guidelines" have, of course, evolved over the years. For history pertinent to the Asilomar discussion, see http://profiles.nlm.nih.gov/FF/Exhibit/narrative/rdna.html (accessed September, 2005).

38. Wade, Chapter 5, "The Conference at Asilomar."

39. Krimsky, 64.

40. Indeed, so concerned was Watson at one point that it is said he threatened a court injunction against a Harvard colleague. Watson's later views on this topic would certainly change to a much more accommodating view. Krimsky, 81.

41. Perhaps J. Lyon and P. Gorner in *Altered Fates: Gene Therapy and the Retooling of Human Life* (New York: W. W. Norton & Company, 1995), 50–52 tell this story best.

42. For a good discussion about the strengths and weakness of the NIH guidelines see Steve Olson, *Biotechnology: An Industry Comes of Age* (Washington, DC: National Academy Press, 1986), 69ff.

43. Lyon and Gorner, 145.

44. Ibid., 161.

45. In the case of fiction over fact, there is David M. Rorvik's *In His Image* (Philadelphia: J. B. Lippincott, 1978). Rorvik argued that he had assisted in the cloning of a human being, presented it as fact, cited the foremost authorities in the field, and collected the rewards of royalties. The book was a fake. When this was discovered, Rorvik said he had wanted to blur the distinction between fact and fiction to provoke public discussion. The story is retold more completely in Jon Tuney, *Frankenstein's Footsteps: Science, Genetics, and Popular Culture* (New Haven, CT: Yale University Press, 1998), 211.

46. The story is told in Lear, 166–171.

47. Interviews conducted in December, 2004.

48. Krimsky, 88.

49. Ibid., 242. This paragraph relies heavily on Krimsky, 242–246.

50. The NIH guidelines were salvaged, of course, and Senator Walter Mondale's recommendation for a National Advisory Commission on the Health Sciences and Society a few years later helped placate both sides. See also Fletcher, 25.

51. Genentech was the first corporation to produce genetic engineering technology for commercial purposes. It will certainly not be the last. See Bankston, 262–263.

52. This should not sound more (or less) ominous than it is. All of the developed countries that are doing research in this area have some form of legislative control over it. Whether it is enough is the subject of controversy. See Simon R. Warne, "The Safety of Work with Genetically Modified Viruses," in *Genetically Engineered Viruses,* ed. Christopher J. A. Ring and Edward D. Blair, New York: Bios Scientific Publications Ltd., 2001, 268ff. For a table of legislation by country, see Deryck Beyleveld and Shaun Pattison, "Legal Regulation of Assisted Procreation, Genetic Diagnosis, and Gene Therapy," in *The Ethics of Genetics in Human Procreation,* ed. Hille Haker and Deryck Beyleveld, Burlington, VT: Ashgate, 2000, 230–233. See also Harold Schmeck, Jr., "Recombinant DNA Controversy: The Right to Know—And to Worry," in *The Gene-Splicing Wars,* ed. Raymond A. Zilinskas and Burke K. Zimmerman, New York: Macmillan Publishing Co., 1986, 93–108 for some hair-raising concerns.

53. The agreed-upon date for the beginning of gene therapy commenced in 1980.

54. United States House of Representatives. *Biotechnology and the Ethics of Cloning: How Far Should We Go? Hearing before the Committee on Science Subcommittee on Technology.* 105th Congress, 1st Session, March 5, 1997. Washington, DC: Government Printing Office, 1997, 2.

55. Suzuki and Knudtson, 193–206.

56. United States House of Representatives, 5.

57. Bernard. D. Davis. "Limits to Genetic Intervention in Humans: Somatic and Germline." In *Human Genetic Information: Science, Law and Ethics,* Ciba Foundation Symposium 149, New York: John Wiley & Sons, 1990, 81. See also French Anderson, "Uses and Abuses of Human Gene Therapy," in *Genetics and Society,* ed. Penelope Barker, New York: H. W. Wilson Co., 1995, 113.

58. May, 76.

59. Ibid.

60. See the tragic story of one scientist's pursuit of Lou Gehrig's disease (ALS) and its unhappy outcome in Jonathan Weiner *His Brother's Keeper,* New York: HarperCollins, 2004. With gene therapy 20 years old when the pursuit began in 2000, it ended in very nearly the same place three year's later with the death of the scientist's brother.

61. The full story is told in Glenn McGee, *Beyond Genetics: Putting the Power of DNA to Work in Your Life* (New York: Harper Collins, 2003), chapter 5. See also Reilly, Philip R, *Abraham Lincoln's DNA and Other Adventures in Genetics* (New York: Cold Spring Harbor Laboratory Press, 2000), 256–260.

62. Michael Boylan and Kevin E. Brown, *Genetic Engineering: Science and Ethics on the New Frontier* (Upper Saddle River, NJ: Prentice Hall, 2001), 127.

63. This distinction is outlined in virtually every citation given so far. See, for example, G.J.V. Nossal and Ross L. Coppel, *Reshaping Life: Key Issues in Genetic Engineering* (New York: Cambridge University Press, 2002), 120–181. But it is not universally accepted as uniform, which leads to further confusion. See Michele Boiani and Hans R. Scholer, "Determinants of Pluripotency in Mammals," in *Principles of Cloning,* ed. Jose Cibell, et al., Boston: Academic Press, 2002, 109–145.

64. Boiani and Scholer, 122–123.

65. Finn Bowring, *Science and Cyborgs: Biotechnology and the Appropriation of Life* (New York: Verso, 2003), 42.

66. R. Michael Blaese. "Germ-Line Modifications in Clinical Medicine." In *Designing Our Descendants: The Promises and Perils of Genetic Modifications,* ed. Audrey R. Chapman and Mark S. Frankel. Baltimore: Johns Hopkins Press, 2003, 75.

67. Boiani and Scholer, 122–123. See also Robert P. Lanza, "Xenotrasplantation," in *Understanding Cloning,* ed. Sandy Fritz, New York: Scientific American, 2002, 84–85.

68. Ibid., 124–125.

69. Jeremy Rifkin, *Algeny* (New York: The Viking Press, 1983), 11.

70. Boiani and Scholer, 180–181.

71. Ibid.

72. Kurt Bayertz, *GenEthics: Technological Intervention in Human Reproduction as Philosophical Problem* (New York: Cambridge University Press, 187), 1.

73. See Chapter 1 of this book and Lee M. Silver, *Remaking Eden: How Genetic Engineering and Cloning Will Transform the American Family* (New York: Avon Books, 1998), 123.

74. Moshe Sipper and James Reggia. "Go Forth and Replicate." In *Understanding Cloning,* ed. Sandy Fritz. New York: Scientific American, 2002, 119–121.

75. President George W. Bush, "Remarks by the President on Stem Cell Research," The Bush Ranch, Crawford, TX, August 9, 2001, www.whitehouse.gov/news/release s/2001/08/20010809-2.html (accessed January, 2005).

76. For example, David McConnell, "The Machine in Man," in *Designing Life? Genetics, Procreation and Ethics,* ed. Maureen Junker-Kenny, Brookfield, VT: Ashgate, 1999, 54ff; Bill McKibben, *Enough: Staying Human in an Engineered Age,* New York: Henry Holt and Company, 2003; Gordon Graham, *Genes: A Philosophical Inquiry,* New York: Routledge, 2002; *Ethical Issues in the New Genetics: Are Genes Us?* ed. Brenda Almond and Michael Parker, Burlington, VT: Ashgate, 2003; *Designing Our Descendants: The Promises and Perils of Genetic Modifications,* ed. Audrey R. Chapman and Mark S. Frankel, Baltimore, Johns Hopkins Press, 2003; Laurence E. Karp, MD, *Genetic Engineering: Threat or Promise?* Chicago: Nelson-Hall, 1976; and Leon R. Kass, "New Beginnings in Life," in *The New Genetics and the Future of Man,* ed. Michael P. Hamilton Grand Rapids, MI: William B. Eerdmans, 1972.

77. Hwa A. Lim, *Genetically Yours: Bioforming, Biopharming, Biofarming* (New Jersey: World Scientific, 2002), 172.

78. One of the better discussions appears in Matti Harry, "Deeply Felt Disgust— A Devlinian Objection to Cloning Humans," in *Ethical Issues in the New Genetics: Are Genes Us?* ed. Brenda Almond and Michael Parker, Burlington, VT: Ashgate, 2003, 61ff.

79. Fletcher, 28–30. Needless to say, this author strongly favors so-called positive eugenics.

80. Robert H. Blank, *The Political Implications of Human Genetic Technology* (Boulder, CO: Westview Press, 1981), 27.

81. One of the more comprehensive surveys, with an assessment of each genetic prospect, is by John C. Avise in *The Hope, Hype & Reality of Genetic Engineering,* New York: Oxford University Press, 2004. It should be pointed out that Avise is very much agreeable to most (if not all) genetic engineering and so is not an impartial critic. His view of a procedure that does not bode well for the future is generally predicated on its financial prospects combined with its funding possibilities. Occasionally he will make an ethical statement, but these are reserved for mostly the easy cases.

82. Peter R. Wheale and Ruth M. McNally, *Genetic Engineering: Catastrophe of Utopia?* (New York: St. Martin's Press, 1988), 243–245.

83. Reilly, 250ff.

84. Lila Guterman. "Teaching the Body to Kill Cancer." *The Chronicle of Higher Education* (November 12, 2004): A22–A26 (accessed via www.chronicle.com, November, 2004; password required).

85. We generally think of a phrase like this with disgust, thinking it refers to the elitist millionaires and billionaires. While this is certainly true, and even in the following

example, it's important to remember that it also applies to those who wrote songs of protest. Consider, for example, the case of David Crosby of the rock group Crosby, Stills, Nash, and Young. While vehemently protesting Vietnam and former President Nixon in the 1960s and 1970, Mr. Crosby appears to have cashed in on his later fame in his sixties when a liver transplant he needed came in record time.

86. Anderson, 113.

87. Suzuki and Knudston, 181–184.

88. Ibid., 255; Jeremy Rifkin, *Algeny, sic passim;* Kass, 42–44; Ira H. Cameron, *Cloning and the Constitution: An Inquiry into Governmental Policymaking and Genetic Experimentation* (Madison WI: Wisconsin University Press, 1985), 6–10; Stephen S. Hall, *Invisible Frontier: The Race to Synthesize a Human Gene* (New York: Atlantic Monthly Press, 1987), 53–60, 240–265.

89. "Send in the Clones," *U.S. News & World Report* 131, no. 7 (August 20, 2001): 12. He promised to do this in a year, which, of course, did not happen.

90. R. Baird Shuman. "Cloning." In *Encyclopedia of Genetics,* ed. Jeffrey A. Knight. Vol. 1. Pasadena, CA: Salem Press, 1999, 112–113.

91. Charles Krauthammer. "Of Headless Mice... and Men." In *Biotechnology,* New York: H. W. Wilson Co., 2000, 97ff.

92. Boylan and Brown, 132–133. One Randolfe Wicker is desirous of cloning himself and was delighted to learn in 2004 that movement in that direction was improving. See Nell Boyce, "The Clone Is Out of the Bottle," *U.S. News & World Report* 136, no. 7 (February 23, 2004): 40 (accessed via *InfoTrac Expanded Academic ASAP Plus,* September, 2004.

93. Colin Tudge. "Growing Pains." *New Statesman* 127, no, 4411 (November 13, 1998): 34 (accessed via *InfoTrac Expanded Academic ASAP Plus,* September, 2004).

94. Henry I. Miller and Gregory Conko. "Let a Hundred Gene-Spliced Flowers Bloom." *The Chronicle of Higher Education,* http://chronicle.com/weekly/v50/i40/40b01201.htm (accessed August, 2004).

95. Megan Garvey. "California Stem Cell Project Energizes Other States to Act." *New York Times* (November 22, 2004).

96. Surprisingly, these were predictions Kass made in his "New Beginnings in Life" in *The New Genetics and the Future of Man,*" ed. Michael P. Hamilton, Grand Rapids, MI: William B. Eerdman Publishing Co., 1972, 44–45 long before they surfaced to become the major ethical issues today.

97. Boylan and Brown, 134, 142. The authors contend that no woman or scientists would do this "because it would be unethical." This strikes one (perhaps the cynical among us) as naive. We have on record scientists plagiarizing or fabricating research for not only much less money but something as illusive as "fame."

98. For more about this, see Tom Bethell, "Mengele Medicine," *American Spectator* 37, no. 9 (November, 2004): 34–35.

99. Raymond A. Zilinskas. "Recombinant DNA Research and Biological Warfare." In *The Gene-Splicing Wars,* ed. Raymond A. Zilinskas and Burke K. Zimmerman. New York: Macmillan Publishing Co., 1986, 195–199. Note the date of this book, nearly 15 years before the September 11, 2001 catastrophe!

100. Julie Ann Miller. "Lessons from Asilomar." *Science News* 127 (February 23, 1985): 122–125 (accessed via *Info Trac Expanded Academic ASAP Plus,* September, 2004).

101. For Roszak, Kass, 154.

CHAPTER 4

Frankenfood or Miracle-Gro? Agricultural Applications of Genetic Engineering

1. Michael J. Reiss and Roger Straughan, *Improving Nature? The Science and Ethics of Genetic Engineering* (New York: Cambridge University Press, 1996), 2–3.

2. Rob De Salle and David Lindley, *The Science of Jurassic Park and the Lost World* (New York: HarperCollins, 1997), 65–70.

3. Reiss, 131.

4. Martin Teitel and Kimberly A. Wilson *Genetically Engineered Food: Changing the Nature of Nature* (Rochester, VT: Park Street Press, 2001), 1.

5. Ibid., 5.

6. Hennery J. Miller and Stephen J. Ackerman, "Perspective on Food Biotechnology" *FDA Consumer* (March 1990) Vol. 24 (2), 12 (accessed via *Info Trac's Expanded Academic ASAP Plus,* October, 2004).

7. Ibid., 9.

8. Brian Tokar, *Redesigning Life?* (New York, Zed Books, 2001), 8.

9. Ibid. Soy beans, cotton, and corn have nearly all been modified in some manner. See also P. R. Reilly, *Abraham Lincoln's DNA and Other Adventures in Genetics* (New York: Cold Spring Harbor Laboratory Press, 2000), 158.

10. Ibid., 17.

11. Ricarda A. Steinbrecher. "Ecological Consequences of Genetic Engineering." In Brian Tokar, *Redesigning Life?* New York, Zed Books, 2001, 74–75.

12. E. Yoxen, *The Gene Business: Who Should Control Biotechnology* (New York: Harper & Row, 1983), 3–5. Some argue that even if found completely safe, do we want plants to produce milk instead of cows?

13. David Suzuki and Peter Knudston, *GenEthics: The Clash Between the New Genetics and Human Values* (Cambridge, MA: Harvard University Press, 1989), 313; Bill Lambrecht, *Dinner at the New Gene Café: How Genetic Engineering Is Changing What We Eat, How We Live, and the Global Politics of Food* (New York: St. Martin's Press, 2001), 43–45.

14. The story about the monarch butterfly has been widely reported. What has not is the evidence since acquired that indicates the butterfly was unaffected by the *Bt* grain after all. See Pinstrup-Andersen, Per and Ebshioler, *Seeds of Contention: World Hunger and the Global Controversy over GM Crops* (Baltimore: Johns Hopkins University Press, 2001), 1. See also the end of this chapter.

15. Britt Bailey, *Engineering the Farm: Ethical and Social Aspects of Agricultural Biotechnology,* ed. Britt Bailey and Marc Lappé. Washington, DC: Island Press, 2002),

xvi–xix. For rBSt and an additional note on *Bt*, see *Genetic Engineering: Opposing Viewpoints* ed. James D. Torr, ed. Greenhaven Press: San Diego, 2001, 13–14.

16. Sheldon Krimsky. "Ethical Issues Involving Production, Planting and Distribution of Genetically Modified Crops." In *Engineering the Farm: Ethical and Social Aspects of Agricultural Biotechnology*, ed. Britt Bailey and Marc Lappé. Washington, DC: Island Press, 2002, 12–13.

17. Reilly, 164. In the case of Europe, dockyard workers refused to unload soybeans (or anything with soy in it) that had been genetically modified, whether they were labeled so or not. Europeans first coined the term Frankenfoods and it has since received widespread acceptance.

18. Quoted in Teitel and Wilson, 11.

19. Paul B. Thompson. "Why Biotechnology Needs and Opt Out." In *Engineering the Farm: Ethical and Social Aspects of Agricultural Biotechnology*, ed. Britt Bailey and Marc Lappé. Washington, DC: Island Press, 2002, 32–34.

20. Lambrecht, 3.

21. Brewster Kneen. "Biotechnology." In *Engineering the Farm: Ethical and Social Aspects of Agricultural Biotechnology*, ed. Britt Bailey and Marc Lappé. Washington, DC: Island Press, 2002, 53.

22. Reiss and Straughan, 158.

23. Daniel J. Kevles. "Diamond v. Chakrabarty and Beyond: The Political Economy of Patenting Life." in *Private Science: Biotechnology and the Rise of the Molecular Sciences*, ed. Arnold Thackeray. Philadelphia: University of Pennsylvania Press, 1998, 65–72. I follow Kevles' account very closely unless otherwise noted. See also Bruce Schacter, *Issues and Dilemmas of Biotechnology* (Westport, CT: Greenwood Press, 1999), 25–33.

24. William Boyd. "Wonderful Potencies." In *Engineering Trouble: Biotechnology and Its Discontents*, ed. Rachael D. Schurman, Denis Doyle, and Takahaski Kelso. Berkeley: University of California Press, 2003, 38–39.

25. A. Chakrabarty. "Patenting of Life Forms." In *Who Owns Life?* ed. David Magnus, Arthur Capalan, and Geleen McGee. Amherst, NY: Prometheus Books, 2002, 18.

26. Jack Wilson. "Patenting Organisms." In *Who Owns Life?* ed. David Magnus, Arthur Capalan, and Geleen McGee. Amherst, NY: Prometheus Books, 2002, 26–27.

27. Boyd, 41.

28. Wilson, 29.

29. Schacter, 32.

30. Ibid., 36.

31. Ibid., 47–53.

32. Boyd, 52.

33. One of the best discussion of this is David Magnus, "Intellectual Property and Agricultural Biotechnology: Bioprospecting or Biopiracy?" In *Who Owns Life?* ed. David Magnus. Arthur Capalan, and Geleen McGee, Amherst, NY: Prometheus Books, 2002, 265–267.

34. J. H. Dodds, *Plant Genetic Engineering* (New York: Cambridge University Press, 1985), 1.

35. S. H. Mantell, et al., *Principles of Plant Biotechnology* (Oxford, London: Blackwell Scientific Publications, 1985), 3.

36. William Bains, *Genetic Engineering for Almost Everybody* (New York: Penguin Books, 1987), 8–9.

37. A form of this is quoted in Teitel and Wilson, 6. The question uppermost (addressed later) on inquiring minds is why one would want to cross a tree with a carrot in the first place.

38. June Goodfield, *Playing God* (New York: Random House, 1977), 36.

39. Rachael D. Schurman. "Biotechnology in the New Millennium." In *Engineering Trouble: Biotechnology and Its Discontents*, ed. Rachael D. Schurman, Denis Doyle, and Takahaski Kelso. Berkeley: University of California Press, 1.

40. R. Hull. "Viruses as Vectors for Plant Genes." In J. H. Dodds, *Plant Genetic Engineering*, New York: Cambridge University Press, 1985, 95.

41. Charles Piller and Keith R. Yamamoto, *Genetic Wars: Military Control over the New Genetic Technologies* (New York: Basic Books, 1988), 185–188. Unless otherwise noted, I have followed this account closely. See also Reiss and Straughan, 116–117.

42. Pat Spallone, *Generation Games: Genetic Engineering and the Future of Our Lives* (Philadelphia: Temple University Press, 1992), 79.

43. Ibid.

44. Francis Fukiyama, *Our Posthuman Future: Consequences of the Biotechnological Revolution* (New York: Farrar, Straus and Giroux, 2002), 196–197.

45. Richard J. Mahoney. "The United States Should Continue Investing in Biotechnology." In *Genetic Engineering: Opposing Viewpoints,* ed. Carol Wekesser. San Diego: Greenhaven Press, 1996, 30–31.

46. Belinda Martineau, *The Creation of the Flavr Savr Tomato and the Birth of Genetically Engineered Food* (New York: McGraw-Hill, 2001). This is the best and most comprehensive history of the story of the Flavr Savr Tomato. Martineau was involved in the development, however, so I have used other research in retelling this story for the sake of balance.

47. Reiss and Straughan, 132, and John Dyson, "Genetic Engineering Improves Agriculture." In *Genetic Engineering: Opposing Viewpoints,* ed. Carol Wekesser, San Diego: Greenhaven Press, 1996, 68–70.

48. Elizabeth Marshall, *High-Tech Harvest: A Look at Genetically Engineered Foods* (New York: Franklin Watts, 1999), 16–17.

49. Keith Redenbaugh et al. "Determination of the Safety of Genetically Modified Crops." In *Genetically Modified Foods*, ed. Karl-Heinz Engel et al. Washington, DC: American Chemical Society, n. d., 76.

50. Dodds, 4. Statistics cited in the next sentence are also from this source.

51. Ibid., and 158.

52. Christopher Wills, *The Wisdom of Genes: New Pathway in Evolution* (New York: Basic Books, 1989), 115. There must be some tongue-in-cheek meant here as some of Wills' work reads like a scene out of *The Fly.*

53. Sheldon Krimsky, *Biotechnology & Society: The Rise of Industrial Genetics* (New York: Praeger, 1991), 92.

54. Sue Hubbell, *Shrinking the Cat: Genetic Engineering Before We Knew About Genes* (Boston: Houghton Mifflin Co., 2001), 18–19.

55. Ibid., 20.

56. Mae-Wan Ho, *Genetic Engineering: Dream or Nightmare?* (New York: Continuum, 2000), 6.

57. David Barling. "The European Response to GM Foods: Rethinking Food Governance." In *Engineering the Farm*, ed. Britt Bailey and Marc Lappé. Washington, DC: Island Press, 2002, 96.

58. Ibid., 97; Maarten J. Chrispeels and David E. Sadava, *Plants, Genes, and Biotechnology*, 2nd ed. (Sudbury, MA: Jones and Bartlett, 2001), xvii.

59. Dean D. Metcalfe. "Allergenicity of Foods Produced by Genetic Modification." In *Genetically Modified Crops: Assessing Safety*, ed. Keith T. Atherton. New York: Taylor & Francis, 2002, 95. Some of the problems are respiratory complications, eye irritation, skin rashes, and gastrointestinal muscosa and submuscosa. This, as we shall see later in this chapter, remains hotly debated. New research in late 2005 appears for now to explain the allergy and GM foods connection, however.

60. Nick Tomlinson. "The Regulatory Requirements for Novel Foods." In *Genetically Modified Crops: Assessing Safety*, ed. Keith T. Atherton. New York: Taylor & Francis, 2002, 45, 55, 61.

61. Barling, 101.

62. Martineau, 240–241.

63. Lambrecht, 219–223; 232–327. See especially the photograph in Lambrecht, 315: a tomato with the head of a fish!

64. "Biotechnology: Europeans Still Wary of Frankenstein Food." *European Report* (March 19, 2003): 462. Also, "Why The French Say 'Non Merci' to Agricultural Biotech," *On the Plate* (April 26, 2003); Felicia Wu, "Perceptions of Food that Are an Ocean Apart," *Financial Times* (May 13, 2004): 15; "Public Uninformed, Unconcerned about Bioengineered Food," *Medical Letter on the CDC & FDA* (December 16, 2001): 17; Kenneth Klee et al. "The Big Food Fight: Europeans Are Railing against 'Frankenstein Foods'—Genetically Modified Crops that Abound in America. And Exporters Have Been Forced to Listen." *Newsweek International* (September 13, 1999) (accessed via *Infotrac Expanded Academic ASAP Plus*, October, 2004). Saba Anna, Anna Moles, and Lynn J. Frewer,. "Public Concerns about General and Specific Applications of Genetic Engineering: A Comparative Study between the UK and Italy." *Nutrition & Food Science* Number 1 (January/February 1998): 19–29.

65. Herbert Gottweis, *Governing Molecules: The Discursive Politics of Genetic Engineering in Europe and the United States* (Cambridge: MA: MIT Press, 1998), 246–247.

66. Glenn McGee, *Beyond Genetics: Putting the Power of DNA to Work in Your Life* (New York: HarperCollins, 2003), 136.

67. David Goodman, Bernado Sorj, and John Wilkinson, *From Farming to Biotechnology: A Theory of Agro-Industrial Development* (New York: Basil Blackwell, 1987), 140. The authors take quite an optimistic view of these developments while admitting some difficulties in ironing out troublesome details.

68. Thomas Shannon, *What Are They Saying about Genetic Engineering* (New York: Paulist Press, 1985), 75.

69. Reiss and Straughan, 127–130.

70. BST or BGH milk is controversial to some because such cows are also treated with IGF-1, an insulin growth hormone to prevent infections. IGF-1 has been linked to human breast and gastrointestinal cancers. See Teitel and Wilson, 34–36.

71. Lynn J. Frewer, Chaya Howard, and Richard Shepherd. "Genetic Engineering and Food: What Determines Consumer Acceptance." *British Food Journal* 9, no. 8 (1995): 31–36.

72. James Freeman. "You're Eating Genetically Modified Food." In *Biotechnology*, New York: H. W. Wilson Co., 2000, 160ff.

73. Ronnie Cummins and Ben Lilliston, *Genetically Engineered Food: A Self-Defense Guide for Consumers* (New York: Marlowe & Company, 2000), 1–6.

74. E. Nestmann et al. "The Regulatory and Science-based Safety Evaluation of Genetically Modified Food Crops—A USA Perspective." In *Genetically Modified Crops: Assessing Safety*, ed. Keith A. Atherton. New York: Taylor & Francis, 2002, 17–21.

75. Vandana Shiva, *Tomorrow's Biodiversity* (Thanes & Hudson, 2000), 18–19.

76. Cummins and Lilliston, 74.

77. Teitel and Wilson, 5.

78. Ibid., 18.

79. Glodie Blumenstyk. "A New Johnny Appleseed." *Chronicle of Higher Education* at http://chronicle.com/weekly/v49/i45/45a02501.htm (accessed October, 2004).

80. Teitel and Wilson, 142–143.

81. Cummins and Lilliston, 108–109.

82. Shiva, "Genetically Engineered 'Vitamin A Rice.'". In *Redesigning Life?* ed. Brian Tokar. New York: Zed Books, 2001, 40–41; Mahabub Hossain. "Opening Address: The Challenge to Feed the World's Poor." *Improving Yield, Stress Tolerance, and Grain Quality*, Novartis Foundation, Symposium 236. New York: John Wiley & Sons, 2001, 1–9.

83. Chrispeels and Sadava, 164.

84. Jennifer Ferrar and Micahel K. Dorsey. "Genetically Engineered Foods: A Minefield of Safety Hazards." In *Redesigning Life?* ed. Brian Tokar. New York: Zed Books, 2001, 62–63; Karl-Heinz Engel. "Foods and Food Ingredients Produced via Recombinant DNA Techniques." In *Genetically Modified Foods*, ed. Karl Heinz Engle et al. Washington, DC: American Chemical Society, n. d., 2–3.

85. Peter Rosset. "Taking Seriously the Claim that Genetic Engineering Could End Hunger." In *Engineering the Farm*, ed. Britt Bailey and Marc Lappé. Washington, DC: Island Press, 2002, 81–87; Ronald Bailey. "Genetically Modifying Food Crops Is Ethical." In *The Ethics of Genetic Engineering*, ed. Lisa Yount. San Diego: Greenhaven Press, 2002, 34–36; Robert Paarlberg. "Promoting Genetically Modified Crops in Developing Countries Is Ethical." In *The Ethics of Genetic Engineering*, ed. Lisa Yount. San Diego: Greenhaven Press, 2002, 52–54; Lambrecht, 273–275; Rev. Michael Oluwatuyi. "How Will We Feed Africa?" and Jordan J. Ballor. "A Theological Framework for Evaluation Genetically Modified Food." *Acton Institute for the Study of Religion & Liberty*, www.acton.org/newsletters/enviromental/articles/09-17-04_oluwatuyi.html

present compelling arguments for GM foods from the Christian perspective (accessed December 2004).

86. Ricarda A. Steinbrecher. "Ecological Consequences of Genetic Engineering." In *Redesigning Life?* ed., Brian Tokar. New York: Zed Books, 2001, 88–89; B. J. Barkla, R. Vera-Estrella, and O. Pantoja. "Towards the Production of Salt-Tolerant Crops." In *Chemicals via Higher Plant Bioengineering*, ed. Fereidoon Shahidi et al. New York: Kluwer Academic/Plenum Publishers, 1999, 77–102; Kazuko Yamaguchi-Shinozaki and Kazuo Shinozaki. "Improving Plant Drought, Salt and Freezing Tolerance by Gene Transfer of a Single Stress-Inducible Transcription Factor." In *Rice Biotechnology: Improving Yield, Stress, Tolerance, and Grain Quality.* Novartis Foundation, Symposium 236, ed. J. A. Goode and D. Chadwick. New York: John Wiley & Sons, 2001, 176–178; Chrispeels and Sadava, 375.

87. E. Nestmann et al., 3–17; Helen Kreuzer and Adrianne Massey. *Recombinant DNA and Biotechnology: A Guide for Students*, 2nd ed. (Washington, DC: ASM Press, 2001), 293–294.

88. Chrispeels and Sadava, 362. This is based on the 20 most important food crops grown in each region and domesticated there.

89. M.G.K. Jones. "Applications of Genetic Engineering to Agriculture." In *Plant Genetic Engineering*, ed. J. H. Dodd. New York: Cambridge University Press, 1985, 260–263; Steve Olson, *Biotechnology: An Industry Comes of Age* (Washington, DC: National Academy Press, 1986), 22–23.

90. Fred Edwords. "Genetic Engineering Can Be Ethical." In *The Ethics of Genetic Engineering*, ed. Lisa Yount. San Diego: Greenhaven Press, 2002, 11–14; "Bioluminescence: The Glowing Tobacco Plant" at http://library.thinkquest.org/18258/noframes/tobacco.htm (accessed April, 2005); Michael D. Lemonick. "Of Fireflies and Tobacco Plants" at www.Time.com/time/Archives/pr eview/0,10987, 962873,00.html (accessed April, 2005).

91. John C. Avise, *The Hope, Hype, and Reality of Genetic Engineering* (New York: Oxford University Press, 2004), 24–25.

92. Reilly, 158–169. Michael Glueck and Robert J. Cihak. "Biotech Foods Foolishly Feared by Franken-Folk." *Jewish World Review* (July 16, 2004) www.jewishworldreview.com/0704/medicine.men071604.asp (accessed October, 2004); Serena Lei. "Franken-Foods No Scarier then the Rest, Report Says" www.medillnewsdc.com/cgi-bin/ultimatebb.cgi?ubb=get_topic&f=7&t=000211 (accessed October, 2004); "In Defense of Demon Seed." *The Economist (US)* 347, no. 8072 (June 13, 1998): 13–15 (accessed via *Infotrac's Expanded Academic Access ASAP Plus*, November, 2004).

93. Chrispeels and Sadava, 530–538.

94. Leigh Ann Williams, "A History of Great Escapes," *Time International* 159, no. 19, (May 20, 2002): 56 (accessed via *InfoTrac's Expanded Academic ASAP Plus*, November, 2004). All four examples are from this source.

95. Luke Anderson, *Genetic Engineering, Food, and Our Environment* (White River Junction, VT: Chelsea Green Publishing, 1999), 54–55; Finn Bowring, *Science and Cyborgs: Biotechnology and the Appropriation of Life* (New York: Verso, 2003), 112–113; Derrick A. Purdue, *Anti-GentiX: The Emergence of the Anti-GM Movement*

(Aldershot, Hampshire, UK: Ashgate, 2000), 4–5, where Purdue sees this loss of biodiversity as potentially cataclysmic.

96. Anne K. Hollander. "Genetic Engineering Will Not Reduce World Hunger." In *Genetic Engineering: Opposing Viewpoints*, ed. David Bender et al. San Diego: Greenhaven Press, 1990, 154–158.

97. Suzanne Havala Hobbs. "Engineered Foods Losing Their Luster." (accessed via www.onthetable.net, December, 2004); Oscar L. Frick. "The Potential for Allergenicity in Transgenic Foods." In *Genetically Modified Foods*, ed. Karl-Heintz Engle et al. Washington, DC: American Chemical Society Press, n. d., 100–112; Jane Rissler and Margaret Mellon, *The Ecological Risks of Engineered Crops* (Cambridge, MA: The MIT Press, 1996), 56–58; Lambrecht, 10–12; 276–280; Mae-Wan Ho, *Genetic Engineering: Dream or Nightmare?* (New York: Continuum, 2000), 18–22.

98. Martha Crouch, "From Golden Rice to Terminator Technology." In *Redesigning Life?* ed. Brian Tokar. New York: Zed Books, 2001, 30. The grass example following is also from here. Ho, 154, lists seven dangers of the instability of transgenic lines. Stephen Nottingham, *Genescapes: The Ecology of Genetic Engineering* (New York: Zed Books, 2002), 166, argues that the benefits of golden rice for eyesight were exaggerated.

99. Orin Langelle. "From Native Forests to Frankenforest." In *Redesigning Life?* ed. Brian Tokar. New York: ZedBooks, 2001, 142–144; Rissler and Mellon, 4–8; 27–29.

100. Ralph Nader. "Genetically Modifying Food Crops Is Not Ethical." In *The Ethics of Genetic Engineering*, ed. Lisa Yount. San Diego: Greenhaven Press, Inc, 2002, 48; Jeremy Rifkin, *The Biotech Century: Harnessing the Gene and Remaking the World* (New York: Putnam, 1998), 2–4; Jonathan Hughes. "Genetically Modified Crops and the Precautionary Principle: Is There a Case for a Moratorium?. In *Ethical Issues in the New Genetics: Are Genes Us?* ed. Brenda Almond and Michael Parker. Burlington, VT: Ashgate, 2003, 146–149; Nottingham, 174–179, lists 20 risks with transgenic crops and plants.

101. Lambrecht, 7.

102. Norman. C. Ellstrand. "When Transgenes Wander, Should We Worry?. In *Engineering the Farm,* ed. Britt Bailey and Marc Lappé. Washington, DC, 2002, 64–66; Miguel A. Altieri, *Genetic Engineering in Agriculture* (Oakland, CA: LPC Group, 2001), 23–25; Rissler and Mellon, 76–78, 87–89; Margaret Mellon. "An Environmental Perspective." In *The Genetic Revolution: Scientific Prospects and Public Perceptions*, ed. Bernard Davis. Baltimore: Johns Hopkins Press, 1991, 64–66.

103. Lambrecht, 106–125.

104. Ibid., 151–155, 206–210.

CHAPTER 5

Well, Hello, Dolly: Animal Applications of Genetic Engineering

1. Unless otherwise noted, I have followed Ian Wilmut's *The Second Creation: Dolly and the Age of Biological Control* (New York: Farrar, Strauss and Giroux, 2000).

See also Gina Kolata, *Clone: The Road to Dolly, and the Path Ahead* (New York: William Morrow and Company), 1998.

2. Phillip Reilly, *Abraham Lincoln's DNA and Other Adventures in Genetics* (New York: Cold Spring Harbor Laboratory Press, 2000), 295.

3. Wilmut, 3.

4. Ibid, 22–23.

5. Ibid., 126–130.

6. Michael J. Reiss and Roger Straughan, *Improving Nature? The Science and Ethics of Genetic Engineering* (New York: Cambridge University Press, 1996), 167.

7. See www.findclinicalstudy.com/d_home_tc.cfm?did=753&bif=p&type=DO& cid=171612 (accessed February, 2005).

8. Ian Wilmut. "The Limits of Cloning." *NPQ: New Perspectives Quarterly* 21 no. 4 (Fall, 2004): 67–73 (accessed via *Academic Search Premier,* January, 2005).

9. Wilmut, 141–145.

10. Ibid., 155.

11. Louis-Marie Houdebine, *Animal Transgenesis and Cloning* (Chichester, UK: John Wylie & Sons, 2003), 3–5.

12. Wilmut, 184ff.

13. Kolata, 30.

14. Wilmut, 216–217.

15. Richard Hull. "The Benefits of Cloning Outweigh the Risks." In *Biomedical Ethics: Opposing Viewpoints,* ed. Leone Bruno et al. San Diego, CA: Greenhaven Press, Inc., 1998, 19.

16. Wilmut, 245–247.

17. Ibid., 267.

18. Seed, who is thought by some to be a bit full of himself, began by promising to clone himself. Later he decided on cloning his wife since the egoism of the idea of cloning himself became too obvious, even to him.

19. For those who missed the movie, the lead character multiplies himself like Xerox copies.

20. A. J. Klotzo. "The Debate about Dolly." In *Cloning,* ed. Michail Ruse and Aryne Sheppard. Amherst, NY: Prometheus Books, 2001, 24.

21. National Bioethics Advisory Commission. "The Risks of Human Cloning Outweighs the Benefits." In *Biomedical Ethics: Opposing Viewpoints,* ed. Leone Bruno, et al. San Diego, CA: Greenhaven Press, Inc., 1998, 26–28.

22. Kolata, 90–92.

23. Wesley J. Smith, *Consumer's Guide to a Brave New World* (San Francisco: Encounter Books, 2004), 52.

24. Many have discussed this book. Kolata does, 93–95.

25. According to Smith, Chung, on her CNN show, wanted to call Rael "Your Holiness." This event, coupled with her disastrous reporting during the Oklahoma City bombing of the Murrah building, put an end to her career.

26. Smith, 50.

27. Jennifer Cunningham. "Animal-to-Human Organ Transplants." In *Biomedical Ethics: Opposing Viewpoints,* ed. Leone Bruno, et al. San Diego, CA: Greenhaven Press, Inc., 1998, 88–90.

28. Norman Maclean. "Transgenic Animals in Perspective." In *Animals with Novel Genes,* ed. Norman Maclean. New York: Cambridge University Press, n. d., 3–5.

29. See various chapters of *Animals with Novel Genes,* ed. Normal Maclean (New York: Cambridge University Press, n. d.). Individual chapters on transgenic insects, fish, birds, rodents, and mammals make for informative reading. Also, Yann Echelard and Gary M. Meade, "Protein Production in Transgenic Animals." In *Comprehensive Biochemistry,* ed. G. Bernardi. Vol. 38. Boston: EIC Laboratories, 2003, 630–632; "Texas A&M College of Veterinary Medicine Clones First Deer." *DVM News Magazine* (August, 2004), 38 (accessed via *Academic Search Premier,* January, 2005).

30. John C. Avise, *The Hope, Hype & Reality of Genetic Engineering* (New York: Oxford University Press, 2004), 85.

31. Richard J. Reece, *Analysis of Genes and Genomes* (West Sussex, England: John Wiley & Sons, Ltd., 2004), 362–363.

32. Ibid., 363–364.

33. Edward J. Alcamo, *DNA Technology: The Awesome Skill* (New York: Harcourt Academic Press, 2001), 280. Fetal and stem cells are discussed on p. 283.

34. Bernice Schacter, *Issue and Dilemmas of Biotechnology* (Westport, CT: Greenwood Press, 1999), 362–363.

35. Houdebine, 18.

36. Joseph Levin and David Suzuki, *The Secret of Life* (Boston: WGBH, 1993), 188–189. This technique allows for the creation of mutant mice for research.

37. John Sedivy and Alexandra L. Joyner, *Gene Targeting* (New York: W. H. Freeman and Co., 1992), 174.

38. Ibid., 132.

39. Houdebine, 156–157; Tim Stewart. "Genetic Modification of Animals." In *Exploring Genetic Mechanisms,* eds. Maxine Singer and Paul Berg. Sausalito, CA: University Science Books, 1997, 590–592; Norman Maclean. "Transgenic Animals in Perspective." In *Animals with Novel Genes,* ed. Norman MacLean. New York: Cambridge University Press, n. d., 13. Some of the diseases listed above, such as Tay-Sachs, Lesch-Nyhan, cystic fibrosis, and Fabry disease are targeted for gene therapy or cloned genes but have had targeting problems to date.

40. Craig Holdrege, *Genetics and the Manipulation of Life* (Hudson, NY: Lindisfarne Press, 1996).

41. For an interesting discussion of this problem see, among many others, Lee, 194ff. Also, Alcamo, 273. The so-called knockout mouse is another such animal whose genes for a single organ, or organ system, have been knocked out.

42. Alcamo, 271.

43. Avise, 108–109.

44. Houdebine, 168–169

45. Ibid., 178–179. For cost, National Research Council, *Animal Biotechnology: Science-Based Concerns* (Washington, DC: The National Academic Press, 2002), 104–105.

46. Jennifer Ferrar and Michael K. Dorsey. "Genetically Engineered Foods: A Minefield of Safety Hazards." In *Redefining Life,* ed. Brian Tokar. New York: Zed Books, 2001, 53–55.

47. Ibid. Canada has resisted BGH-milk for more than a decade.

48. Bernice Schacter, *Issues and Dilemmas of Biotechnology* (Westport, CT: Greenwood Press, 1999), 63ff.

49. Avise, 80–82.

50. Ibid., 86–88.

51. Alcamo, 275.

52. For example, Hwa A. Lim, *Genetically Yours: Bioinforming and Biopharming, Biofarming* (New Jersey: World Scientific, 2002), 115.

53. For more, see "S. Korea Unveils First Dog Clone" (Wednesday, August 3, 2005) *BBC News, World Edition* at http://news.bbc.co.uk/2/hi/science/nature/4742453.stm (accessed September, 2005).

54. Michael W. Fox, *Superpigs and Wondercorn: The Brave New World of Technology* (New York: Lyons & Burford, 1992), 114–129, 166–170; Reiss and Straughan, 179.

55. Joseph Levine and David Suzuki, *The Secret of Life* (Boston: WGBH), 177. The flowers were "whiter than white." Gene insertions make it possible for the flower industry to create trendy colors and rush them to market. While this seems innocent enough, similar research on animals might raise strong ethical concerns.

CHAPTER 6
Where No Man (or Woman) Has Gone Before: The Human Genome Project

1. This chapter treats only the Human Genome Project. The short length of this book does not allow greater depth for the coverage of other important projects announced seven years after HGP, in 1997. The Human Genome Diversity Project (HGDP) is designed to give us a better understanding of the origin and evolution of humans. The Cancer Genome Anatomy Project, begun the same year, is an effort to catalog all the genes expressed in cancer cells. The Environmental Gene Project attempts to pinpoint so-called environmental risk genes. For more, see I. Edward Alcamo, *DNA Technology: The Awesome Skill* (New York: Harcourt Academic Press, 2001), 322–325. The date given here is the date of the completion of the "working draft" DNA sequence of the human genome. The Human Genome Project began in 1990 and completed all its stated goals (see later this chapter) in 2003 when the 20,000 to 25,000 human genes were identified. Although all the original goals have been met, the HGP will continue for many more years, as data analysis will take decades. This data analysis is still, sometimes confusingly, referred to as the Human Genome Project.

2. Merrill Goozner, *The $800 Million Pill: The Truth behind the Cost of New Drugs* (Los Angeles: University of California Press, 2004), 61.

3. James Shreeve, *The Genome War: How Craig Venter Tried to Capture the Code of Life and Save the World* (New York: Alfred A. Knopf, 2004), 13–104. The story is so compellingly told and so incredibly rent with strife that it seems a miracle the project ever succeeded at all, much less with such fanfare and triumph. See especially, 3–9.

4. Interestingly, in the 2004 elections presidential contender John Kerry claimed that funding was insufficient. John Leo, "Stem Cell Debate Is Riddled with Dishonesty," *Conservative Chronicle* (August 25, 2004): 2.

5. Ronald J. Trent, "Milestones in the Human Genome Project: Genesis to Postgenome," *MJA* 173 (December 2000): 591.

6. Ingrid Wickelgren, *The Gene Masters: How a New Breed of Scientific Entrepreneurs Raced for the Biggest Prize in Biology* (New York: Time Books, 2002), 90ff. The early history of the Human Genome Project is told in about four chapters in Wickelgren and is a must-read.

7. Meredith Wadman, "Biology's Bad Boy," *Fortune* 149, no. 5 (March 8, 2004): 166–173 (accessed via *EBSCOhost,* February, 2005, subscription required).

8. Glenn McGee, *The Perfect Baby: A Pragmatic Approach to Genetics* (Lanham, MD: Rowman & Littlefield Publishers, 1997), 17.

9. Ibid.

10. Ingrid Wickelgren, *The Gene Masters: How a New Breed of Scientific Entrepreneurs Raced for the Biggest Prize in Biology* (New York: Time Books, 2002), 11.

11. Malcolm Ritter, "Scientists Cut Estimate of Human Gene Count," *Charlotte Observer* (Thursday, October 21, 2004): 4A.

12. W. Gilbert, "Human Genome Sequencing," in *Biotechnology and the Human Genome: Innovations and Impact,* ed. Avril D. Woodhead (New York: Plenum Press, 1988), 29.

13. Hwa A. Lim, *Genetically Yours: Bioinforming, Biopharming, Biofarming* (Hackensack, NJ: World Scientific, 2000), 4. All figures number base pairs.

14. Francis S. Collins, "The Human Genome Project and the Future of Medicine," in *Great Issues for Medicine in the Twenty-First Century,* ed. Dana Cook Grossman and Heinz Valtin (New York: The Academy of Sciences, 1999), 42–44.

15. Quoted in Craig Holdrege, *Genetics and the Manipulation of Life* (Hudson, NY: 1996), 82.

16. Bowring, 147.

17. Márcio Fabri Dos Anjos, "Power, Ethics and the Poor in Human Genetic Research," in *The Ethics of Genetic Engineering* (New York: Orbis Books, 1998), 80–81.

18. Phillip R. Sloan, "Completing the Tree of Descartes," in *Controlling Our Destinies,* ed. Phillip R. Sloan (Notre Dame, IN: University of Notre Dame Press, 2000), 10–16.

19. Timothy Lenoir and Marguerite Hays, "The Manhattan Project for Biomedicine," in *Controlling Our Destinies,* ed. Phillip R. Sloan (Notre Dame, IN: University of Notre Dame Press, 2000), 29–62. The Manhattan Project refers to secret work on the development of the atomic bomb during World War II.

20. Alice Domurat Dreger, "Metaphors of Morality in the Human Genome Project," in *Controlling Our Destinies,* ed. Phillip R. Sloan (Notre Dame, IN: University of Notre Dame Press, 2000), 180–186.

21. For example, knowing you have a marker for a certain disease, should you have children? Or, knowing that your prenatal child has a certain marker for a fatal disease, do you abort? The problem lies with the markers. Markers only make broad predictions; they are not destiny.

22. S. J. Fitzgerald and Kevin T. Fitzgerald, "Philosophical Anthropologies and the HGP," in *Controlling Our Destinies,* ed. Phillip R. Sloan (Notre Dame, IN: University of Notre Dame Press, 2000), 405–407; for "big money," see Daniel E. Koshland Jr., "Sequences and Consequences of the Human Genome," in *Emerging Issues in Biomedical Policy,* vol. 1, ed. Robert H. Blank and Andrea L. Bonnicksen (New York: Columbia University Press, 1999), 178.

23. Fred Sanger, "The Early Days of DNA Sequencing," *Nature Medicine* 7, no. 1 (March 2001): 268.

24. Karl Drlica, *Understanding DNA and Genome Cloning* (New York: John Wiley & Sons, Inc., 1997), 272–273.

25. Jeremy Rifkin, *The Biotech Century: Harnessing the Gene and Remaking the World* (New York: Jeremy P. Tarcher/Putnam, 1998), 190–191; for the body's ten trillion cells, see Graeme Laurie, *Genetic Privacy: A Challenge to Medico-Legal Norms* (New York: Cambridge University Press, 2002), 86, note 5.

26. Richard J. Reece, *Analysis of Genes and Genomes* (West Sussex, England: John Wiley & Sons, Ltd, 2004), 294–295.

27. These are publicly available at http://www.ncbi.nlm.nih.gov/dbEST (accessed February, 2005).

28. Ibid., 308. Recall that the HGP involves more than humans alone.

29. Scientists are divided over whether to refer to DNA as a code. For 50 years or more this has been the vernacular, but many scientists today prefer talking about it as a language and the decoding as if translating that language. Codes must be broken, at which point they are no longer useful. Languages, once translated, are useful forever after.

30. A similar example is found in Gregory Stock and John Campbell, eds., *Engineering the Human Germline: An Exploration of the Science and Ethics of Altering the Genes We Pass to Our Children* (Oxford, UK: Oxford University Press, 2000), 21. The enjambment reads as follows: "like this, but backwards."

31. Quoted in British Medical Association, *Our Genetic Future: The Science and Ethics of Genetic Technology* (Oxford, UK: Oxford University Press, 1992), 132.

32. Suzuki and Knudtson, 110; Paul Debenham, "The Use of Genetic Markers for Personal Identification in the Analysis of Family Relationships," in *Human Genetic Information: Science, Law and Ethics* (New York: John Wiley & Sons, 1990), 37–39; Norman H. Crawley and P. E. Crawley, "Commercial Exploitation of the Human Genome: What Are the Problems?" In *Human Genetic Information: Science, Law and Ethics* (New York: John Wiley & Sons, 1990), 133–135. For band and gene probes, see Alcamo, 293–295.

33. Michael Fortun, "The Human Genome Project and the Acceleration of Biotechnology," in *Private Science: Biotechnology ands the Rise of the Molecular Sciences,* ed. Arnold Thackery (Philadelphia: University of Philadelphia Press, 1998), 182–183 and 186.

34. Nossal and Coppel, 42–48.

35. Rick J. Carlson and Gary Stimeling, *The Terrible Gift: The Brave New World of Genetic Medicine* (New York: Public Affairs, 2002), especially 96–100.

36. Schacter, 75–80.

37. Shreeve, 27–29, on science and egos.

CHAPTER 7

The Doctor Will See You Now: Genetic Engineering and the Treatment of Diseases

1. J. Lyon and P. Gorner, *Altered Fates: Gene Therapy and the Retooling of Human Life* (New York: W. W. Norton, 1995), 28. Berg's $125-million Beckman Center for Molecular and Genetic Medicine is counting on this opinion.

2. Geoff Scott, "A Century of Medical Miracles," *Current Health 2, A Weekly Reader Publication* 18, no. 5 (January 1922): 4–10 (accessed via *Expanded Academic ASAP Plus*, December, 2004).

3. Carl Feldbaum, "Some History Should Be Repeated," *Science* 296 No. 5557 (8 February 2002): 975.

4. Jeffery Saver and Titi Tamburi, "Genetics of Cerebrovascular Disease," in *Neurogenetics,* Stefan-M. Puslt (New York: Oxford University Press, 2000), 403.

5. Nell Boyce, "Fat Chance for Meat," *U.S. News & World Report* 136, no. 6 (February 16, 2004): 62–65 (accessed via *academic Search Premier,* January, 2005).

6. Gregory E. Pence, *Recreating Medicine: Ethical Issues at the Frontiers of Medicine* (New York: Rowman & Littlefield Publishers, Inc., 2000), 3–5.

7. Marque-Lusia Miringoff, *The Social Costs of Genetic Welfare* (New Brunswick, NJ: Rutgers University Press, 1991), 51–52.

8. Maxwell J. Mehlman, *Wondergenes: Genetic Enhancement and the Future of Society* (Bloomington: Indiana University Press, 2003), 122–124; I.Q. discussion, Francis Fukiyama, *Our Posthuman Future: Consequences of the Biotechnology Revolution* (New York: Farrar, Straus, and Giroux, 2002), 25–32.

9. Michael Reiss and Roger Straughan, *Improving Nature? The Science and Ethics of Genetic Engineering* (New York: Cambridge University Press, 1996), 102–105.

10. Gregory Stock, *Redesigning Humans: Our Inevitable Genetic Future* (New York: Houghton Mifflin Co., 2002), 83–89. Stock indicates the incentive: if nearly every adult took one for life, at a dollar a day for everyone over 45 this would translate into $30 billion annually in the United States alone. Eric Juengst, et al. "Biogerontology, Anti-Aging Medicine and the Challenges of Human Enhancement," *Hastings Center Report* 33, no. 4 (July-August 2003): 21–30; David Stipp, "This Man Would Have

You Live a Really, Really, Really, Really Long Time," *Fortune* 149, no. 12 (June 16, 2004): 136–141 (accessed via *EBSCOhost* March, 2005, subscription required).

11. Philip Kitcher, "Patients in the 21st Century: The Impact of Predictive Medicine," in *Great Issues for Medicine in the Twenty-First Century*, ed. Dana Cook Grossman and Heinz Valtin (New York: New York Academy of Sciences, 1999), 142.

12. B. J. Barnhart, et al., "The Human Genome Initiative: Issues and Impacts," in *Biotechnology and the Human Genome: Innovations and Impact*, ed. Avril D. Woodhead (New York: Plenum Press, 1988), 103.

13. Jon Turney, *Frankenstein's Footsteps: Science, Genetics, and Popular Culture* (New Haven: Yale University Press, 1998), 8–9, 100, 101.

14. Marc Lappé, *Genetic Politics: The Limits of Biological Control* (New York: Simon and Schuster, 1979), 117.

15. Glenn McGee, *The Perfect Baby: A Pragmatic Approach to Genetics* (Lanham, MA: Rowman & Littlefield Publishers, 1997), 8–10. Of course, Nazi Germany extended the eugenics debate to its horrible conclusion—genocide.

16. Lappé, 119, 137.

17. Bernard Rollin, *The Frankenstein Syndrome: Ethical and Social Issues in the Genetic Engineering of Animals* (New York: Cambridge University Press, 1995), 16–17.

18. Glenn McGee, *Beyond Genetics: Putting the Power of DNA to Work in Your Life* (New York: HarperCollins, 2003), 59.

19. Kitcher, 143.

20. Hope Shand, "Gene Giants: Understanding the "Life Industry," in *Redesigning Life?* ed. Brian Tokar (New York: Zed Books, 2001), 233.

21. Troy Duster, "Hidden Eugenics Potential," in *Designing Our Descendants: The Promises and Perils of Genetic Modifications*, ed. Audrey R. Chapman and Mark S. Frankel (Baltimore: Johns Hopkins University Press, 2003), 161.

22. Luke Anderson, *Genetic Engineering, Food and Our Environment* (White River Junction, VT: Chelsea Green Publishing Co., 1999), 76.

23. Rick Weiss, "U.S. Researchers Reach Deal in '99 Gene Therapy Case," *Washington Post* (February 10, 2005): A03 (accessed February, 2005 via www.washingtonpost.com/wp-dyn/articles/A12136-2005Feb9.html?sub=AR, registration required).

24. Sheldon Krimsky, *Biotechnics & Society: The Rise of Industrial Genetics* (New York: Praeger, 1991), 164–165.

25. "Genetic Engineering Promises a Long Line of Improvements to Animals—From Fish that Glow to Mosquitoes without Disease—But Are Federal Regulators Keeping a Watchful Eye?" (accessed via *EBSCOHost*, February, 2005, subscription required).

26. Jeremy Rifkin, *The Biotech Century: Harnessing the Gene and Remaking the World* (New York: Jeremy P. Tarcher/Putnam, 1998), 48.

27. Some see screening as hidden eugenics at work. Troy Duster, "Hidden Eugenics Potential," in *Designing Our Descendants: The Promises and Perils of Genetic Modifications*, ed. Audrey R. Chapman and Mark S. Frankel (Baltimore: Johns Hopkins University Press, 2003), 160–161.

28. Finn Bowring, *Science, Seeds and Cyborgs: Biotechnology and the Appropriation of Life* (New York: Verso, 2003), 149–150.

29. Ibid., 180; Ainsely Newson, "Is There a Cost in the Choice of Genetic Enhancement," in *Ethical Issues in the New Genetics: Are Genes Us?* ed. Brenda Almond and Michael Parker (Burlington, VT: Ashgate, 2003), 23, 32–33.

30. June Goodman, *Playing God* (New York: Random House, 1977, 59), 104–107. See Chapter 9 for more on this theme.

31. Gilbert S. Omenn, "Genetics and Public Health: Historical Perspectives and Current Challenges and Opportunities," in *Genetic and Public Health in the 21st Century,* ed. Muin J. Khoury, et al. (New York: Oxford University Press, 2000), 38. For diabetes, Andy Coghlan, "UK Cloners Target Diabetes Cure," *New Scientist* 183, no. 2462 (August 21, 2004): 8–10 (accessed via Gale Group, February, 2005, subscription required).

32. Bryan Appleyard, "Are Genes Us?" in *Ethical Issues in the New Genetics: Are Genes Us?* ed. Brenda Almond and Michael Parker (Burlington, VT: Ashgate, 2003), 1.

33. Richard Overy, *The Dictators: Hitler's Germany and Stalin's Russia* (New York: W. W. Norton & Company, 2004), 243–244.

34. Graeme Laurie, *Genetic Privacy: A Challenge to Medico-Legal Norms* (New York: Cambridge University Press, 2002), 102–103.

35. Mae-Wan Ho, *Genetic Engineering: Dream or Nightmare?* (New York: Continuum, 2000), 267.

36. Joseph G. Perpich, "The Recombinant DNA and Bioterrorism," *Chronicle of Higher Education* (March 15, 2002), (accessed January, 2005 via chronicle.com/weekly/v48/i27/27b02001.htm, subscription required).

37. "Stem Cell Facts" at www.house.gov/weldon/Stem/Facts.htm (accessed January, 2005).

38. "Stem Cell," *Wikipedia: The Free Encyclopedia* at http://en.wikipedia.org/wiki/Stem_cell_research#Cord_blood_stem_cells (accessed May, 2005).

39. Of course, former President Reagan began his political career as a Democrat.

40. "The Science of Stem Cells" at www.house.gov/weldon/Stem/Science.htm (accessed January, 2005). Let partisan research already cited in this book confirm what is repeated here.

41. Deryck Beyleveld and Shaun Pattinson, "Legal Regulation of Assisted Procreation, Genetic Diagnosis and Gene Therapy," in *The Ethics of Genetics in Human Procreation* (Burlington, VT: Ashgate, 2000), 235–241.

42. "The Science of Stem Cells."

43. Mary Carmichael, "Medicine Next Level," *Newsweek* 144, no. 23 (December 6, 2004): 44–49 (accessed March, 2005 via *EBSCOhost*, subscription required).

44. Andrew Pollack, "Rare Infection Is Confirmed in 2nd Patient on M. S. Drug," *New York Times* (March 4, 2005) at www.nytimes.com/2005/03/04/health/04drug.html (accessed March, 2005). Officials at Tysabri, the company making the drug, announced the deadly brain infection. Some may think that two are not very many but, with one dead and another with the same infection in clinical trials, the finger is doing more than pointing. The infection, known as progressive multifocal leukoencephalopathy (PML), is so rare that even one case is alarming. Two in the same trial was enough to

shut the company down. While the link with genetic engineering is admittedly weak, it is enough to raise eyebrows on the process, at least in the case of this disease.

45. J. S. Mill is most famously remembered for his utilitarian concern (often criticized) that if 30 people in a room of 31 decided to eat the 31st, utilitarian philosophy would not find this problematic. Needless to say, the 31st person had considerable difficulty with the idea.

46. Bonnie Steinbock, "Ethical Differences between in Heritable Genetic Modification and Embryo Selection," in *Designing Our Descendants: The Promises and Perils of Genetic Modifications,* ed. Audrey R. Chapman and Mark S. Frankel (Baltimore: Johns Hopkins University Press, 2003), 179ff.

47. As in the case of a mother purposely getting pregnant only in order to create a placenta for another son or daughter to use for a disease such as cystic fibrosis. Though such stories can be found in the press, no known disease can be successfully treated with this procedure, though hope springs eternal.

48. Jeffrey Brainard, "Stem Cells that Qualify for Federal Funds May Be Useless for Treatment, Study Says," *Chronicle of Higher Education* 51, no. 22 (February 4, 2005): A22.

49. Silla Brush, "Hoping to Avoid the Brain-Drain, States Push to Finance Stem Cell Research," *Chronicle of Higher Education* 51, no. 22 (February 4, 2005): A22.

50. John Finn, "Stem Cell Hype Is Hard to Combat," *Los Angeles Times* (September 1, 2005). See also Wesley J. Smith, "Science Unstemmed," *American Spectator* (February 2005) 23–26. Smith calls Proposition 71 just another "experiment in Big Biotech's lab of horrors."

51. J. L. Edwards, et al., "Cloning Adult Farm Animals: A Review of the Possibilities and Problems Associated with Somatic Cell Nuclear Transfer," *American Journal of Reproductive Immunology* 50 (2003): 113–123.

52. Ibid.

53. Fred Charles Iklé, "The Deconstruction of Death: The Coming Politics of Biotechnology," *The National Interest* (Winter 2000–2001): 87–95 (accessed via *Academic Search Premier,* January, 2005).

54. Jeremy Rifkin, 24; blood vessels and pancreases, Robert S. Boyd, "Bioengineers Create Spare Parts for Body," *Charlotte Observer* (August 26, 2004): 14A.

55. Carolyn Williams, "Human Cloning, Genetic Engineering and Privacy," at http://www.yale.edu/ynhti/curriculum/units/2000/3/00.03.07.x.html (accessed September, 2005).

56. John Harris, *Clone, Genes and Immortality: Ethics and the Genetic Revolution* (New York: Oxford University Press, 1998), 124–127.

57. Nancy Montagne, "Cystic Fibrosis: Identification of the Gene," in *Emerging Issues in Biomedical Policy,* Vol. 1, ed. Robert H. Bland and Andrea L. Bonnicksen (New York: Columbia University, 1992), 182–185.

58. "Combination Stem Cell–Gene Therapy Approach Seen as Potential Treatment for Cystic Fibrosis," at www.stemcellresearchfoundation.org/WhatsNew/December_ 2004.htm#4 (accessed February, 2005).

59. Helen Spencer, "The Potential for Stem Cell Therapy in Cystic Fibrosis," *Journal of the Royal Society of Medicine,* 97, no. 44 (2004): 55 (accessed at www. rsmpress.co.uk/s44–52.pdf, February, 2005).

60. Philip R. Reilly, *Abraham Lincoln's DNA and Other Adventures in Genetics* (New York: Cold Springs Harbor Press, 2000), xvii.

61. Philip R. Reilly, *Genetics, Law and Social Policy* (Cambridge, MA: Harvard University Press, 1977), chapters 2 and 4 especially. See also, Stephen S. Coughlin and Wylie Burke, "Public Health Assessment of Genetic Disposition to Cancer," in *Genetics and Public Health in the 21st Century*, ed. Muin J. Khoury, et al. (New York: Oxford University Press, 2000), 151–160.

62. David Suzuki and Peter Knudtston, *Genethics: The Clash Between the New Genetics and Human Values* (Cambridge, MA: Harvard University Press, 1989), 171–173; Penelope Barker, ed. *Genetics and Society* (New York: H. W. Wilson, Co., 1995), 89–101.

63. See for example, Tony Hunter, "Onocogenes, Growth Suppressor Genes, and Cancer," in *Exploring Genetic Mechanisms,* ed. Maxine Singer and Paul Berg (Sausalito, CA: University Science Books, 1997), chapter 4.

64. J. Ettienne-Decant, *Genetic Biochemistry: From Gene to Protein* (New York: John Wiley & Sons, 1988), 166–170.

65. One such breakthrough involves laryngeal cancer. Scientists successfully used antisense RNA to block that cancer's tumorousness. See Edward I. Alcamo, *DNA Technology: The Awesome Skill* (New York: Harcourt Academic Press, 2001), 135.

66. Ibid.

67. H.M.D. Gurling, "Recent Advances in the Genetics of Psychiatric Disorder," in *Human Genetic Information: Science, Law and Ethics* (New York: John Wiley & Sons, 1990), 48–50; Edwin Cook, Jr., "Genetics of Psychiatric Disorders: Where Have We Been and Where Are We Going?" at ajp.psychiatryonline.org/cgi/content/full/157/7/1039 (accessed February, 2005).

68. "Psychiatric Disorders Detected Using Blood Test," at www.news-medical.net/?id=7390 (accessed February, 2005).

69. "Brain Serotonin Enzyme Finding Might Explain Psychiatric Disorders," at news.mc.duke.edu/news/article.phd?id=7703 (accessed February, 2005).

70. Alcamo, 173.

71. Ibid, 178.

72. Andrew Pollack, "Method to Turn Off Bad Genes Is Set for Tests on Human Eyes," *New York Times* (September 24, 2004) at www.nytimes.com/2004/09/14/business/14gene.html?th (accessed September, 2004).

73. Lakshmi Sandhana, "Chips Coming to a Brain Near You," *Wired News* (October 22, 2004) at www.wired.com/news/archive/0,2618,2004-10-22,00.html (accessed February, 2005).

74. John Avise, *The Hope, Hype, and Reality of Genetic Engineering* (New York: Oxford University Press, 2004), 118–119.

75. Thomas F. Lee, *Gene Future: The Promise and Perils of the New Biology* (New York: Plenum Press, 1993), 153–155. See also "Muscle Stem Cells Show Promise against Muscular Dystrophy in Mouse Model," at www.niams.nih.gov/ne/press/2002/07_03.htm (accessed February, 2005); Yoshihide Suanada, "The Muscular Dystrophies," in *Neurogenetics,* ed. Stefan-M. Puslt (New York: Oxford University Press, 2000), 78–85.

76. Carol Krause, "Genetic Testing Can Save Lives," in *Biomedical Ethics: Opposing Viewpoints,* ed. Bruno Leone, et al. (San Diego, CA: Greenhaven Press, 1998), 202–203.

77. Lyon and Gorner, 7–8.

78. Glenn McGee, *Beyond Genetics: Putting the Power of DNA to Work in York Life* (New York: HarperCollins, 2003), 77.

79. "Genetic Link Made in Breast Cancer," *Charlotte Observer* (February 28, 2005) at www.charlotte.com/mld/observer/news/10967601.htm?1c (accessed February, 2005).

80. Lyon and Gorner, 31.

81. Ibid., 31–33.

82. Ibid., 77.

83. Ibid., 86.

84. Rick J. Carlson and Gary Stimeling, *The Terrible Gift: The Brave New World of Genetic Medicine.* (New York: Public Affairs, 2002), 111–112.

85. Ibid., 215–221. The history is a most useful one to review. Anderson is credited with the birth. At the time of Lyon and Gorner's history, Anderson was involved in 26 of the then-known 37 gene protocols worldwide, p. 281.

86. Reilly (2000), xvi–xvii. Reilly argues that somatic gene therapy, or the correction of disease by delivering normal genes to the cells of affected individuals, "cannot claim a single cure."

87. Carlson and Stimeling, 24–25.

88. Ho, 249.

89. Gregory Stock and John Campbell, *Engineering the Human Germline: An Exploration of the Science and Ethics of Altering the Genes We Pass to Our Children* (Oxford, UK: Oxford University Press, 2000), 46–47.

90. Marcy Darnovsky, "The Case Against Designer Babies: The Politics of Genetic Enhancement," in *Redesigning Life?* ed. Brian Tokar (New York: Zed Books, 2001), 147.

91. Fred Edwards, "Genetic Engineering Can Be Ethical," in *The Ethics of Genetic Engineering,* ed. Lisa Yount (San Diego, CA: Greenhaven Press, Inc.), 16.

92. Michael Hamilton, ed., *The New Genetics and the Future of* Man (Grand Rapids, MI: William B. Eerdmans, 1972), 49–50, emphasis in the original. Also quoted in Jon Turney, *Frankenstein's Footsteps: Science, Genetics, and Popular Culture* (New Haven: Yale University Press, 1998), 160.

93. Karl A. Drlica, *Double-Edged Sword* (New York: Addison-Wesley Publishing, Co. 1994), 12–13.

94. For the latter, whether it is ethical to knowingly bring a child with genetic defects to term, see Shelia A. McLean in *The Ethics of Genetics in Human Procreation,* ed. Hille Haker and Deryck Beyleveld (Aldershot, UK: Ashgate, 2000), 20–25.

95. Rifkin, 30–31.

96. Finn, 234. See also www.galegroup.com/free_resources/whm/trials/babym.htm (accessed March, 2005).

97. Lee M. Silver, *Remaking Eden: How Genetic Engineering and Cloning Will Transform the American Family* (New York: Avon Books, 1998), 98.

98. In the case of *Davis v. Davis* 824 S.W.2d 558 (Tenn. 1992), the embryos were awarded to the mother. See also Melanie Blum, "Embryos and the New Repro-

ductive Technologies" at http://www.surrogacy.com/legals/embryotech.html (accessed September, 2005).

99. "Perfect?" *The Economist* (April 14, 2001) (accessed via *Expanded Academic ASAP Plus,* January, 2005).

100. Frederick Grinnell, "Defining Embryo Death Would Permit Important Research," *Chronicle of Higher Education* (May 16, 2003) at http://chronicle.com/ weekly/v49/i36/36b01301.htm (accessed November, 2004, subscription required).

101. Nicholas Wade, "Tracking the Uncertain Science of Growing Heart Cells," *New York Times* (March 14, 2005) (accessed March, 2005 via www.nytimes.com, registration required).

CHAPTER 8

Just the Facts, Ma'am: Genetic Engineering, DNA Evidence, and the Courts

1. The terms DNA fingerprinting and DNA profiling are actually synonymous. But DNA fingerprinting is most often used in the press, while practitioners refer to profiling. Susan Aldridge, *The Thread of Life: The Story of Genes and Genetic Engineering* (New York: Cambridge University Press, 1996), 162.

2. I have relied on Robert Cook-Deegan, *The Gene Wars: Science, Politics and the Human Genome* (New York: W. W. Norton & Co.), 299–300.

3. DNA testing was first used in American courts in 1986 but did not come full circle until after the Jeffreys case. Cook-Deegan, 302.

4. Ibid.

5. Ibid.; Edward I. Alcamo, *DNA Technology: The Awesome Skill* (New York: Harcourt Academic Press, 2001), 208–217.

6. *American Journal of Law and Medicine* 18, no. 3 (1992): 287; William C. Thompson, "Evaluating the Admissibility of New Genetic Identification Tests: Lessons from the DNA Wars," *The Journal of Criminal Law & Criminology* 84, no. 1 (1993): 22–104; Jordan K. Garrison, "Courts Face the Exciting and the Inevitable: DNA in Civil Trials," *The Review of Litigation* 23, no. 2 (Spring, 2004): 435–461.

7. "When the Evidence Lies: Joyce Gilchrist Helped Send Dozens to Death Row. The Forensic Scientist's Errors Are Putting Capital Punishment under the Microscope," *Time* (May 21, 2001) Vol. 157 (20), 38–41.

8. Karl A. Drlica, *Double-Edged Sword* (New York: Addison-Wesley Publishing Co., 1994), 122.

9. Ibid.

10. Robert S. Boyd, "Frozen Ark to Preserve DNA of Species at Risk," *New York Times* (October 16, 2004):18A.

11. Ronald Bailey, "Unlocking the Cells," in *Biotechnology* (New York: H. W. Wilson Co., 2000), 39. I've greatly simplified the procedure found here.

12. G. V. Nossal and Ross L. Coppel, *Reshaping Life: Key Issues in Genetic Engineering* (New York: Cambridge University Press, 2002), 95.

13. British Medical Association, *Our Genetic Future: The Science and Ethics of Genetic Technology* (Oxford: Oxford University Press, 1992), 206.

14. Aldridge, 163.

15. Michael J. Reiss and Roger Straughan, *Improving Nature? The Science and Ethics of Genetic Engineering* (New York: Cambridge University Press, 1996), 7–8.

16. Bernice Schacter, *Issue and Dilemmas of Biotechnology* (Westport, CT: Greenwood Press, 1999), 146–151. The description of the process is from Schacter but reduced to its simplest parts.

17. Alcamo, 217.

18. Sandy Primrose, Richard Twyman, and Bob Old, *Principles of Gene Manipulation,* 6th edition (Malden, MA: Blackwell Science, 2001), 281; Thomas F. Lee, *Gene Future: The Promise and Perils of the New Biology* (New York: Plenum Press, 1993), 46–47; Patricia A. Ham, "An Army of Suspects: The History and Constitutionality of the U.S. Military's DNA Repository and Its Access for Law Enforcement Purposes," *The Army Lawyer* (July/August 2003):1–19.

19. "O.J. vs. the (1-in-170 Million) Odds," *U.S. News & World Report* 118, no. 20 (May 22, 1995). (Accessed via EBSCOHost's *Academic Search Premier,* April, 2005, subscription required).

20. P. R. Reilly, *Abraham Lincoln's DNA and Other Adventures in Genetics,* (New York: Cold Spring Harbor Laboratory Press, 2000), 55.

21. Quoted in Lee, 48.

22. Ibid., 79. The paper is called "Aggressive Behavior, Mental Sub-Normality and the XYY Male." See also Suzuki and P. Knudston, *Genethics: The Clash between the New Genetics and Human Values* (Cambridge, MA: Harvard University Press, 1989), 145–157.

23. Suzuki and Knudston, 155.

24. Francisco Corte-Real, "Forensic DNA Databases," *Forensic Science International* 146 (December, 2004): S142; "DNA Forensics Work, but the Databanks Need to Be Filled," *Science & Government Report* 39, no. 9 (May 15, 2001): 5 (accessed via *Expanded Academic ASAP Plus,* January, 2005).

25. Reiss and Straughton, 213.

26. Ibid., 197.

27. Reilly, 146–148.

28. Ibid., 147–149. One of the more notorious researchers in question is Dean Hamer, who studied gay men and their siblings. The study proved to be filled with inconsistencies. Further aspersions were cast when Hamer admitted his own gay orientation.

29. Ibid., 147.

30. See "Is there a Gay Gene?" NARTH, National Association for Research and Therapy of Homosexuality, at http://www.narth.com/docs/istheregene.html (accessed September, 2005).

31. Timothy F. Murphy, "Genetic Science and Discrimination," *Chronicle of Higher Education* (June 4, 2004) at www.chronicle.com/weekly/v50/i39/39b01701.htm (accessed July, 2004).

32. Schacter, 145.

33. Jennifer A. Dlouhy, "Opposition Leaves DNA Inmate Access Bill a Missing Link from Evidence Chain," *CQ Weekly* 62, no. 39 (October 9, 2004): 2382.

CHAPTER 9
Endgame: Genetic Engineering, Future Trends, Current Recommendations

1. Alexander Pope, *Rape of the Lock,* ed. Cynthia Wall (Boston: Bedford Books, 1998), Canto I, lines 1–2, 53.

2. B. J. Barnhart, et al., "The Human Genome Initiative: Issues and Impacts," in *Biotechnology and the Human Genome: Innovations and Impacts,* ed. Avril D. Woodhead (New York: Plenum Press, 1988),106.

3. Alan Ryan, "Eugenics and Genetic Manipulation," in *The Genetic Revolution and Human Rights: The Oxford Amnesty Lectures 1998,* ed. Justine Burley (Oxford, UK: Oxford University Press, 1999), 125.

4. The red state/blue state dichotomy came to us via the Gore-Bush campaign and was later solidified by the Bush-Kerry elections. The upshot is that the country is very evenly divided between red states (in which more conservative-minded voters voted for Bush) and blue states (in which more liberal-minded voters voted for Gore). The majority of the US land mass is red, versus much smaller blue land mass (but with much larger populations). The 2004 elections followed a nearly identical division between red and blue states.

5. Kurt Bayertz, *GenEthics: Technological Intervention in Human Reproduction as a Philosophical Problem* (New York: Cambridge University Press, 1987), 23.

6. Quoted in Pat Spallone, *Generation Games: Genetic Engineering and the Future of Our Lives* (Philadelphia: Temple University Press, 1992), 201.

7. Ho, Mae-Wan, *Genetic Engineering: Dream or Nightmare?* (New York: Continuum, 2000), 11.

8. Nicholas Agar, *Liberal Eugenics: In Defense of Human Enhancement* (Malden, MA: Blackwell Publishing, 2004).

9. Ibid., vii.

10. Amitai Etzioni, *Genetic Fix* (New York: Macmillan Publishing Co., 1973), 101; Bernard B. Davis, "Novel Pressures on the Advance of Science," *Ethical and Scientific Issues Posed by Human Uses of Molecular Genetics* (New York: New York Academy of Sciences), 200. Bernard writes, "I would further emphasize a distinction between biomedical technology, which aims at preventing and alleviating illnesses, and the kinds of technology that aim at bigger and better consumption."

11. Ibid., 103.

12. Shattuck, *Forbidden Knowledge* (New York: St. Martin's Press, 1996), *sic passim.*

13. Shattuck, 100. His Chapter VI treats as examples of destructive knowledge the main two here, recombinant DNA and the Human Genome Project.

14. Sheldon Krimsky, *Science in the Private Interest: Has the Lure of Profits Corrupted Biomedical Research?* (New York: Rowman & Littlefield, Inc., 2003), 125–126.

15. Quoted in Krimsky, 129.

16. James Shreeve, *The Genome War: How Craig Venter Tried to Capture the Code of Life and Save the World* (New York: Alfred A. Knopf, 2004). Even after Venter was fired and the company's stock plummeted, his net worth was still in excess of $100 million, down from $500 million. Good men have averted their gazes for far less. This much money will not guarantee an averted gaze but it surely raises the odds. More significant than even this, however, is Shreeve's concentration on the vast egos among the early principal characters, Collins, Venter, and Watson. It does not bode well for the future to have such power in the hands of men so easily swayed by personal aggrandizement.

17. Krimsky, 130–131.

18. Names listed on research often number in the half-dozens because just about anyone can append his or her name to a paper for supplying only one footnote. This is a small exaggeration, but not by much. Scientific papers are notorious for having casts of thousands in the authorship line.

19. Krimsky, 131.

20. Bernard E. Rollin, *The Frankenstein Syndrome: Ethical and Social Issues in the Genetic Engineering of Animals* (New York: Cambridge University Press, 1995), 70. The researcher points to the release of killer bees in California. With a name like that, one would have thought more than mundane care would have been indicated.

21. Ibid., 141.

22. William W. Lowrance, *Modern Science and Human Values* (New York: Oxford University Press, 1985), 6. I am indebted to Tom Moore for pointing this book out to me.

23. Ibid., 7.

24. Wesley J. Smith, *Consumer's Guide to a Brave New World* (San Francisco: Encounter Books, 2004), 100.

25. Lowrance, 15–20; 61; 90–99. I have changed Lowrance's points slightly for this discussion. Lowrance calls the rhetorical arrogance of some scientists' pronouncements as "unconstructive nonsense," 103.

26. C. S. Lewis, *Pilgrim's Regress: An Allegorical Apology for Christianity, Reason and Romanticism* (Grand Rapids, Eerdmans, 1959), 68.

27. W. French Anderson, "Human Gene Therapy: Scientific and Ethical Considerations," in *Ethics, Reproduction and Genetic Control*, ed. Ruth F. Chadwick (New York: Croom Helm, 1987), 157.

28. Bernard Gert, "Genetic Engineering of Humans is Largely Unethical," in *Genetic Engineering: Opposing Viewpoints*, ed. James D. Torr (San Diego, CA: Greenhaven Press, 2001), 69.

29. Karen Lebacqz, "The Ghosts Are on the Wall," in *The Manipulation of Life*, ed. Robert Esbjornson (San Francisco: Harper & Row, 1984), 24–26.

30. Christian Anfinsen, "Bio-Engineering: Short-Term Optimism and Long-Term Risk," in *The Manipulation of Life*, ed. Robert Esbjornson (San Francisco: Harper & Row, 1984), 42, 48. For more on regulation, see "National Policies to Oversee Inheritable Genetic Modifications Research," in *Designing Our Descendants*,

The Promises and Perils of Genetic Modifications, eds. Audrey R. Chapman and Mark
S. Frankel (Baltimore: Johns Hopkins University Press, 2003), chapter 20.

31. Lebacqz, 32–34.

32. Smith, 104–105.

33. Leon R. Kass, "Triumph or Tragedy? The Moral Meaning of Genetic Technology," in *The Ethics of Genetic Engineering,* ed. Lisa Yount (San Diego: Greenhaven Press, Inc., 2002), 79. Kass makes similar points in "New Beginnings in Life," in *The New Genetics and the Future of Man,* ed. Michael P. Hamilton (Grand Rapids, MI: William B. Eerdsman Publishing Co., 1972), 16–17; 20–21; 40–41.

34. President's Council on Bioethics. *Beyond Therapy: Biotechnology and the Pursuit of Happiness* (Washington, DC, 2003), 43.

35. Andy Miah, "Patenting Human DNA," in *Ethical Issues in the New Genetics: Are Genes Us?* eds. Brenda Almond and Michael Parker (Burlington, VT: Ashgate, 2003), 112–113.

36. Kass, 82–87.

37. Merrill Goozner, *The $800 Million Pill: The Truth Behind the Cost of New Drugs* (Los Angeles: University of California Press, 2004), 30.

38. Gilbert Meilaender, "Mastering Our Gen(i)es: When Do We Say No?" *The Christian Century* 107, no. 27 (October 3, 1990): 872–873.

39. Graeme Laurie, *Genetic Privacy: A Challenge to Medico-Legal Norms* (New York: Cambridge University Press, 2002), 25–30.

40. E. V. Kontorovich, "Human Cloning Is Unethical," in *Genetic Engineering: Opposing Viewpoints,* ed. James D. Torr (San Diego, CA: Greenhaven Press, 2001), 102–103.

41. Charles Piller and Keith Yamamoto, *Gene Wars: Military Control over the New Genetic Technologies* (New York: Beech Tree Books, 1998), 34–35.

42. Bernard Davis, "Comments: The Scientific Chapters," in *Genetic Revolution: Scientific Prospects and Public Perceptions,* ed. Bernard Davis (Baltimore: Johns Hopkins University Press, 1991), 263.

43. George Smith, II, *The New Biology: Law, Ethics, and Biotechnology* (New York: Plenum Press, 1989), 46.

44. President's Council on Bioethics, 38.

45. June Goodfield, *Playing God* (New York: Random House, 1977), 118–120.

46. J. Lyons and Peter Gorner *Altered Fates: Gene Therapy and the Retooling of Human Life* (W. W. Norton & Co., 1995), 544.

47. Gregory Stock, *Redesigning Humans: Our Inevitable Genetic Future* (New York: Houghton Mifflin Co., 2002), 178.

48. James Shreeve, "The Other Stem-Cell Debate," *New York Times* (April 10, 2005) at www.nytimes.com/2005/o4/10/magazine/10CHIMERA.html?th&emc=th (accessed April, 2005).

49. Jeffrey Brainard, "A New Kind of Bioethics," *Chronicle of Higher Education* (May 21, 2004) at chronicle.com/weekly/v50/i37/37a02201.htm (accessed February, 2005).

50. President's Council on Bioethics, xvii.

51. Robert Novak, "Stem Cell Vote Swap," *Conservative Chronicle* 20, no. 14 (April 6, 2005): 29.

52. Gina Kolata and Sheryl Gay, "Koreans Report Ease in Cloning for Stem Cells," *New York Times,* May 20, 2005 at http://query.nytimes.com/gst/abstract. html?res = fb0e13fb3f5d0c738eddac0894dd404482 (accessed June, 2005).

53. See David Cyranoski, "Korea's Stem-Cell Stars Dogged by Suspicion of Ethical Breach," *Nature* 2004;429(3) at http://www.nature.com/news/2004/040503/ pf/429003a_pf.html (accessed June, 2005).

54. Deb Riechmann, "Bush Condemns S. Korea Stem Cell Advances" at http:// www.abcnews.go.com/Politics/wireStory?id = 778442 (accessed June, 2005).

55. E-mail communication, John Finn, from the Associated Press, dated June 2, 2005, page D3.

56. For example, John C. Avise, *The Hope, Hype, & Reality of Genetic Engineering* (New York: Oxford University Press, 2004), *sic passim;* John Harris, *Clones, Genes, and Immortality: Ethics and Genetic Revolution* (New York: Oxford University Press, 1998), 196–211; 222–241; *Recombinant DNA and Biotechnology: A Guide for Students,* 2nd ed., eds. Helen Kreuzer and Adrianne Massey (Washington, DC: ASM Press, 2001), 275–280.

57. Andy Coghlan, "Engineering the Therapies of Tomorrow," *New Scientist* 138, no. 1870 (24 April 1993): 26, 28.

58. Sondra Wheeler, "Contingency, Tragedy and the Virtues of Parenting," in *Beyond Cloning: Religion and the Remaking of Humanity* (Harrisburg, PA: Trinity International, 2001), 120.

59. Thomas F. Lee, *Gene Future: The Promise and Perils of the New Biology* (New York: Plenum Press, 1993), 120.

60. Stock, 141. The pregnancy drug, thalidomide, may provide the moral. It's premature use led to a generation in which thousands of babies were born without arms or legs, or with mere buds of the same. While time perfected the use of such drugs (we are now far more abstemious about their use among pregnant women), the cost was as high as any we have ever paid.

61. Rollin, 92.

62. Ibid., 114–120.

63. Steven Milloy, "Science-Politics Tension Dates back Centuries" at www. foxnews.com/story/0,2933,136999,00.html (accessed April, 2005).

64. Allen D. Berhy, "Playing God," in *Genetic Ethics: Do the Ends Justify the Genes?* eds. John R. Kilner, Rebecca D. Pentz, and Frank E. Young (Grand Rapids, MI.: William B. Eerdmans Co., 1997), 60.

65. Ibid., 61.

66. Lewis Thomas, "Early Morning Thoughts of a Chairman after a Conference," in *The Manipulation of Life,* ed. Robert Esbjornson (San Francisco: Harper & Row, 1984), 1.

67. W. French Anderson, "Genetics and Human Malleability," in *Emerging Issues in Biomedical Policy,* Vol. 1, ed. Robert H. Blank and Andrea L. Bonnicksen (New York: Columbia University Press, 1992), 196.

68. Jon W. Gordon, *The Science and Ethics of Engineering the Human Germ Line* (New York.: Wiley-Liss, 2003), 226.

69. Before her tragic death, Ms. Schiavo could swallow and breathe on her own but could not speak, write, talk, or walk. She was bedridden and cared for 24 hours a day. Her husband sought and acquired a court order to have her feeding tube removed but her parents intervened, thinking that she could be rehabilitated with proper care. A state court ordered the tube removed but Congress passed an act forcing the federal courts to intervene. After an appellate court overruled any future intervention, Ms. Schiavo died.

70. Gordon, 229.

71. Gilbert Meilaender makes a similar point.

72. Thomas A. Shannon, *Made in Whose Image? Genetic Engineering and Christian Ethics* (Amherst, New York: n. p. Originally published Atlantic Highlands, NJ: Humanities Press, 1997), 2000, 123–125. I have adapted Shannon's points for use here.

73. G.J.V. Nossal and Ross L. Coppel, *Reshaping Life: Key Issues in Genetic Engineering* (New York: Cambridge University Press, 2002), 191.

74. Paul Flaman, *Genetic Engineering, Christian Values, and Catholic Teaching* (New York: Paulist Press, 2002), 18–21.

75. Randolph E. Schmid, "Study: Cloned Meat, Milk, Nearly the Same," *My Way News* at http://apnews.myway.come/article/20050411/D89DEC9O0.html (accessed April, 2005).

76. Sheldon Krimsky, *Biotechnics & Society: The Rise of Industrial Genetics* (New York: Praeger, 1991), 16.

77. Spallone, 117.

78. Walter Charles Zimmerli, "Who Has the Right to Know the Genetic Constitution of a Particular Person?" in *Human Genetic Information: Science, Law, and Ethics* (New York: John Wiley & Sons, 1990), 100.

79. This may seem impossible on first reading but the Weyco Company in Wisconsin now regularly tests for nicotine and will fire any employee who tests positive. Not only this, but the company reserves the right to fire a person regardless of whether that testing reveals nicotine consumption on or off the job.

80. Spallone, 16–17.

81. Goodfield, 190.

82. Rick J. Carlson and Gary Stimeling, *The Terrible Gift: The Brave New World of Genetic Medicine* (New York: Public Affairs, 2002), 248.

83. Arinzeh won the Presidential Award in 2004, the nation's highest scientific award, for her forward-thinking research in adult stem cell work. See "NJIT Presidential Award Winner Takes Stem Cell Research Another Step" at www.worldhealth. net/p/416.html (accessed April, 2005). The distinction between adult and embryonic stem cell research may be moot if very recent research proves accurate. A new stem cell procedure on mice allows extraction without the destruction of the embryo. For more see Wade, Nicholas, "Stem Cell Test Tried on Mice Saves Embryo." New York Times, October 17, 2005, accessed via www.nytimes.com/2005/10/10/health/17stem.html, October 2005.

84. For more on this see "History of Chemical Warfare and Current Threat" at www.nbc-med.org/SiteContent/MedRef/OnlineRef/FieldManuals/medman/History.htm (accessed April, 2005). Of course, the threat of biochemical use comes

from terrorists and so research to contain and/or prevent it must be ongoing. The so-called first use of these weapons was banned after World War I and the Geneva Protocol of 1925. See also Meryl Nass, "Can Biological, Toxin, and Chemical Warfare be Eliminated?" *Politics and the Life Sciences* (February, 1992): 30–32 at www. anthraxvaccine.org/CV/Pol_Sci_1992.html (accessed April, 2005).

85. Ibid., 277.

86. My point here is that there is no federal group with this power and authority, and there should be. My thunder here is somewhat stolen by the news that this has just recently been strongly urged. See Nicholas Wade, "Group of Scientists Drafts Rules on Ethics for Stem Cell Research," *New York Times* (April 27, 2005) at http://www. nytimes.com/2005/04/27/health/27stem.html? (accessed April, 2005, registration required).

APPENDIX

1. J. S. Mill, "On Nature," in J. S. Mill, *Three Essays on Religion,* 3rd ed., London: Longmans, 1885.

2. For a fuller defense of this claim, see my books *Practical Ethics,* 2nd ed. (Cambridge, UK: Cambridge University Press, 1993) and *Rethinking Life and Death* (New York: St. Martin's Press, 1996).

Selected Bibliography

PREFACE

Chazan, David. "Who Are the Raelians?" *BBC News* 28 December 2002.

Hooper, Rowan. "Genes Blamed for Fickle Female Orgasm." *New Scientist* 8 June 2005.

Koshland, Daniel E. Jr., "Sequences and Consequences of the Human Genome." *Science* 13 October 1989, 189.

CHAPTER 1

Aldridge, Meryl, and Robert Dingwall. "Teleology on Television?: Implicit Models of Evolution in Broadcast Wildlife and Nature Programmes." *European Journal of Communication* 18, no. 4 (Dec. 2003): 435–455.

Bayertz, Kurt. *GenEthics' Technological Intervention in Human Reproduction as a Philosophical Problem*. New York: Cambridge University Press, 1987.

Beckwith, Jo Ann, Timothy Hadlock, and Heather Suffron. "Public Perceptions of Plant Biotechnology—A Focus Group Study." *New Genetics & Society* 22, no.2 (August 2003): 93–109.

Briggs, Laura, and Jodi I. Kelber-Kaye. "'There Is No Unauthorized Breeding in Jurassic Park': Gender and the Uses of Genetics." *NWSA Journal* 12, no. 3 (Fall 2000): 92–113.

Caulfield, Timothy. "Ethics Watch: Public Perceptions and Regulatory Policy." *Nature Reviews Genetics* 3, no. 12 (December 2002): 902.

"Could Dinosaurs Really Live Again?" *Current Science* 79, no. 2 (17 September 93): 4–7.

Dans, Peter. *Doctors in the Movies: Boil the Water and Just Say Aah.* Bloomington, IL: Medi-Ed Press, 2000.

Fincham, J.R.S. "Beyond Good and Evil." *Nature* 356, no. 6366 (19 March 1992): 203–205.

Gonder, Patrick. "Like a Monstrous Jigsaw Puzzle: Genetics and Race in Horror Films of the 1950s." *Velvet Light Trap: A Critical Journal of Film & Television* Fall 2003; no. 52: 33–45.

Gorke, Alexander, and Georg Ruhrmann. "Public Communication between Facts and Fictions: On the Construction of Genetic Risk." *Public Understanding of Science* 12, no. 3 (2003).

Hettema, Joop, Kees C. Leidelmeijer, and Rinie Geenen. "Dimensions of Information Processing: Physiological Reactions to Motion Pictures." *European Journal of Personality* 14, no. 1 (January/February 2000): 39–63.

Jeffreys, Mark. "Dr. Daedalus and His Minotaur: Mythic Warnings about Genetic Engineering, from J.B.S. Haldane, François Jacob, and Andrew Niccol's Gattaca." *Journal of Medical Humanities* 22, no. 2 (Summer 2001): 137–152.

Kaiser, Jocelyn. "Spielberg Movie to Aid Mouse Research." *Science* 276, no. 5313 (2 May 1997): 667.

Korts, Külliki, Sue Weldon, and Margrét Lilja Guðmundsdóttir. "Genetic Databases and Public Attitudes: A Comparison of Iceland, Estonia and the UK." *Trames: A Journal of the Humanities & Social Sciences* 8, no. 1/2 (2004): 131–149.

Mawer, Simon. "DNA and The Meaning of Life." *Nature Genetics* 33, no. 4 (April 2003): 453–455.

Marinucci, Ron. "Our Genes/Our Choices Who Gets to Know? Genetics and Privacy/Making Better Babies: Genetics and Reproduction/Genes on Trial: Genetics, Behavior, and the Law" (Film). *Library Media Connection* 22, no. 5 (Feb2004): 93.

Schneider, Keith. "Public of 2 Minds on Genetic Shifts." *New York Times.* Late city final edition (31 May 1987): 21, col. 1, sec. 1, part 1.

Skal, David J. *Screams of Reason: Mad Science and Modern Culture.* New York: W. W. Norton & Company, 1998.

Urbanek, Drew, and Bette-Lee Fox. "Genes on Trial: Genetics, Behavior and the Law; Making Better Babies: Genetics and Reproduction; Who Gets To Know: Genetics and Privacy." *Library Journal* 129, no. 9 (15 May 2004):127.

Vergano, Dan, and Susan Wloszczyna. "Genetics at Play in Films." *USA Today* (18 June 2002).

www.answers.com/topic/list-of-mad-scientists

www.epinions.com/Movies-Foreign_ Films-keyword-Mad_Doctor

www.filmsite.org/horrorfilms.html

www.savagecinema.com/drbutcher.htm

CHAPTER 2

Alcamo, I. Edward. *DNA Technology: The Awesome Skill.* New York: Harcourt Press, 2001.

British Medical Association. *Our Genetic Future: The Science and Ethics of Genetic Technology.* Oxford: Oxford University Press, 1992.

Choi, Charles. "Francis Crick, 1916–2004." *Scientific American* 291, no. 4 (October 2004): 41.

Cohen, Philip. "Master & Commander." *New Scientist* 184, no. 2475 (27 November 2004): 36–39.

Coman, Daniel, and Irma M. Russu. "Base Pair Opening in Three DNA-unwinding Elements." *Journal of Biological Chemistry* 280, no. 21 (27 May 2005): 20216–20221.

Davis, Tinsley H. "Meselson and Stahl: The Art of DNA Replication." *Proceedings of the National Academy of Sciences of the United States of America* 101, no. 52 (28 December 2004): 17895–17896. "The Discovery of DNA." *Free Inquiry* 25, no. 2 (February /March 2005): 58.

Drlica, Karl A. *Double-Edged Sword: The Promises and Risks of the Genetic Revolution.* Reading, MA: Addison-Wesley Publishing Company, Helix Books, 1994.

Du, Quan, Chiam Smith, Nahum Shiffeidrim, Maria Vologodskaia, and Alexander Vologodskii. "Cyclization of Short DNA Fragments and Bending Fluctuations of the Double Helix." *Proceedings of the National Academy of Sciences of the United States of America* 102, no. 15 (12 April 2005):5397–5402.

Kevles, Daniel J. *In the Name of Eugenics: Genetics and the Uses of Human Heredity.* Cambridge, MA: Harvard University Press, 1995.

"King of Codes." *Wilson Quarterly* 29, no. 1 (Winter 2005): 100–101.

Lee, Keekok. *Philosophy and Revolutions in Genetics: Deep Science and Deep Technology.* New York: Macmillan, 2003.

Levine, J., and D. Suzuki. *The Secret of Life: Redesigning the Living World.* Boston: WGBH, 1993.

Leslie, Mitch. "After the Double Helix." *Science* Vol. 307, no. 5713 (25 February 2005): 1177.

Lyon, J., and P. Gorner. *Altered Fates: Gene Therapy and the Retooling of Human Life.* New York: W. W. Norton & Co., 1995.

Macphail, Theresa Marie. "The Viral Gene: An Undead Metaphor Recoding Life." *Science as Culture* 13, no. 3 (September 2004): 325–345.

Maienschein, Jane. *Whose View of Life? Embryos, Cloning and Stem Cells.* Cambridge, MA: Harvard University Press, 2003.

Mao, Chengde. "The Emergence of Complexity: Lessons from DNA." *PLoS Biology* 2, no. 12 (December 2004): 2036–2038.

Newnham, David. "Describing DNA." *Times Educational Supplement,* no. 4622 (18 February 2005): 5.

Pyle, Anna Marie. "DNA Repair: Big Engine Finds Small Breaks." *Nature* 432, no. 7014 (11 November 2004): 157–158.

Reilly, P. R. *Abraham Lincoln's DNA and Other Adventures in Genetics.* New York: Cold Spring Harbor Laboratory Press, 2000.

Sato, Yu-ichi, et al. "The Effect of Backbone Structure on Polycation Comb-Type Copolymer/DNA Interactions and the Molecular Assembly of DNA." *Biomaterials* 26, no. 7 (March 2005): 703–711.

Seeman, Nadrian C. "From Genes to Machines: DNA Nanomechanical Devices." *Trends in Biochemical Sciences* 30, no. 3. Regular edition (March 2005): 119–125.
———. "Nanotechnology and the Double Helix." *Scientific American* 290, no. 6 (June 2004): 64–73.
Smith, Wesley J. "Science Unstemmed." *American Spectator* (February 2005) 23–26.
Wexler, Barbara. *Genetics and Genetic Engineering.* New York: Thomson-Gale, 2004.
www.library.thinkquest.org/18258/history.htm
www.bio.davidson.edu/people/ kahales/301Genetics/timeline.html
www.eugenics.net/
www.eugenicsarchive.org/eugenics/
www.eugenics-watch.com/
www.genetics.org/
www.kumc.edu/gec/
www.mbg.cornell.edu/Genetics_History.cfm
www.mendelweb.org/
www.ornl.gov/sci/techresources/Human_Genome/genetics.shtml

CHAPTER 3

Almond, Brenda, and Michael Parker, eds. *Ethical Issues in the New Genetics: Are Genes Us?* Burlington, VT: Ashgate, 2003.
Avise, John C. *The Hope, Hype & Reality of Genetic Engineering.* New York: Oxford University Press, 2004.
Brownlee, Christen. "Perfect Match." *Science News* 167, no. 21 (21 May 2005): 323.
Brooke, James. "Without Apology, Leaping Ahead in Cloning." *New York Times.* Late edition-final (31 May 2005): F1, col. 2.
"Catholics Brace for State Legislative Battles." *America* 192, no. 4 (7 February 2005): 6.
Chapman, Audrey R., and Mark S. Frankel, eds. *Designing Our Descendants: The Promises and Perils of Genetic Modifications.* Baltimore, Johns Hopkins University Press, 2003.
"Clone Rangers Give the Hard Cell." *Times Higher Education Supplement,* no. 1692 (20 May 2005): 4.
"Cloning Pioneer Plans to Open Stem-Cell Bank." *Wall Street Journal* 245, no.107. Eastern edition (2 June 2005): D3.
Duncan, David Ewing. "Bio Brain Backs Stem Cells." *Discover* 26, no. 4 (April 2005): 18–19.
Graham, Gordon. *Genes: A Philosophical Inquiry.* New York: Routledge, 2002.
Gurnham, David. "The Mysteries of Human Dignity and the Brave New World of Human Cloning." *Social & Legal Studies* 14, no. 2 (June 2005): 197–214.
Hansen, Brian. "How Reproductive and Therapeutic Cloning Differ." *CQ Researcher* 14, no. 37 (22 October 2004): 880.
Jeffrey, Terence R. "No Human Monkeys—for Now." *Human Events* 61, no. 17 (16 May 2005): 4.
Kakutani, Michiko. "The Frankenstein Model, but Updated with Cloning." *New York Times* 154, no. 53189 (19 April 2005): 6, The Arts.

Kostel, Ken. "Stem Cell Researchers Move Closer to Cloning Us." *Discover* 26, no.1 (January 2005): 44.

Krimsky, Sheldon. *Genetic Alchemy: The Social History of the Recombinant DNA Controversy.* Cambridge, MA: MIT Press, 1982.

Lear, John. *Recombinant DNA: The Untold Story.* New York: Crown Publishers, 1978.

Lord, Richard. "Whose View of Life? Embryos, Cloning, and Stem Cells." *Science Teacher* 72, no. 1 (January 2005): 72–73.

Lyon, J., and P. Gorner. *Altered Fates: Gene Therapy and the Retooling of Human Life.* New York: W. W. Norton & Company, 1995.

McGee, Glenn. *Beyond Genetics: Putting the Power of DNA to Work in Your Life.* New York: HarperCollins, 2003.

McKibben, Bill. *Enough: Staying Human in an Engineered Age.* New York: Henry Holt and Company, 2003.

Medina, John. "Fox Got Your Tongue?" *Psychiatric Times* 22, no.6 (May2005): 42–45.

Monastersky, Richard. "Researchers Generate Human Stem Cells by Cloning Embryos." *Chronicle of Higher Education* 51, no. 38, (27 May 2005): A15.

Paarlberg, Robert L. "The Great Stem Cell Race." *Foreign Policy* no.148 (May/June2005): 44–51.

Park, Alice, Christine Gorman, Helen Gibson, and Laura A. Locke. "Inside the Korean Cloning Lab." *Time* 165, no. 22 (30 May 2005): 54–57.

Singer, Maxine, and Paul Berg, eds. *Exploring Genetic Mechanisms.* Sausalito, CA: University Science Books, 1997.

Smith, Wesley J. "Cold Utopia." *National Review* 57, no. 11(20 June 2005): 54–55.

Vogel, Gretchen. "Cloning of Human Stem Cells Speeds Up." *Science Now* (19 May 2005): 1–2.

Watman, Max. "Ignorant Armies Clash by Night." *New Criterion* 23, no. 9 (May 2005): 61–67.

Zimmerman, Burke K. *Biofuture: Confronting the Genetic Era.* New York: Plenum Press, 1984.

gslc.genetics.utah.edu/units/cloning/

www.bioscience.org/news/scientis/cloning.htm

www.cloninginformation.org/

www.globalchange.com/clonlink.htm

www.ornl.gov/sci/techresources/Human_Genome/elsi/cloning.shtml

www.srtp.org.uk/cloning.shtml

www.usfca.edu/cloning/

CHAPTER 4

Avise, John C. *The Hope, Hype, and Reality of Genetic Engineering.* New York: Oxford University Press, 2004.

Bailey, Britt, and Marc Lappé, eds. *Engineering the Farm: Ethical and Social Aspects of Agricultural Biotechnology.* Washington, DC: Island Press, 2002.

"Biotech Bans: A Mixed Crop." *California Journal* 35, no. 12 (December 2004): 5.

Boyer, Paul. "Unwarranted Fear of GMOs Harms Us All." *NPQ: New Perspectives Quarterly* 21, no. 4, (Fall 2004): 105–107.

Bren, Linda. "Genetic Engineering: The Future of Foods?" *FDA Consumer* 37, no. 6 (November/ December 2003): 28–34.

"'Frankenfoods' Take Over Grocery Stores." *New York Amsterdam News* 92, no. 15 (12 April 2001): 20.

Fukiyama, Francis. *Our Posthuman Future: Consequences of the Biotechnological Revolution.* New York: Farrar, Straus and Giroux, 2002.

"Our Genetically Modified Future." *Nutrition Health Review: The Consumer's Medical Journal* no. 84 (2002): 3–5.

Hileman, Bette. "Clashes Over Agbiotech." *Chemical & Engineering News* 81, no. 23 (9 June 2003): 25–31.

———. "Europe's Distaste For Agbiotech." *Chemical & Engineering News* 80, no. 28 (15 July 2002): 7.

Marshall, Elizabeth. *High-Tech Harvest: A Look at Genetically Engineered Foods.* New York: Franklin Watts, 1999.

Martineau, Belinda. *The Creation of the Flavr Savr Tomato and the Birth of Genetically Engineered Food.* New York: McGraw-Hill, 2001.

McGee, Glenn. *Beyond Genetics: Putting the Power of DNA to Work in Your Life.* New York: HarperCollins, 2003.

Pelletier, David L. "Science, Law, and Politics in FDA's Genetically Engineered Foods Policy: Scientific Concerns and Uncertainties." *Nutrition Reviews* 63, no. 6 (June 2005): 210–223.

Piller, Charles and Keith R. Yamamoto. *Genetic Wars: Military Control over the New Genetic Technologies.* New York: Basic Books, 1988.

Pollack, Andrew. "Modified Seeds Found Amid Unmodified Crops." *New York Times* 153, no. 52769 (24 February 2004): C6.

Reiss, Michael J. and Roger Straughan. *Improving Nature? The Science and Ethics of Genetic Engineering.* New York: Cambridge University Press, 1996.

Schmidt, Jennifer, et al. "Health Professionals Hold Positive Attitudes Toward Biotechnology and Genetically Engineered Foods." *Journal of Environmental Health* 67, no. 10 (June 2005): 44–49.

Seaman, Donna. "Eating Science: Genetically Modified Foods." *Booklist* 98, no. 7 (1 December 2001): 616.

Shiva, Vandana. *Tomorrow's Biodiversity.* London: Thames & Hudson, 2000.

"Should the FDA Adopt a Stricter Policy on Genetically Engineered Foods? CON." *Congressional Digest* 80, no. 3 (March 2001): 77–86.

Teitel, Martin and Kimberly A. Wilson. *Genetically Engineered Food: Changing the Nature of Nature.* Rochester, VT: Park Street Press, 2001.

Thackeray, Arnold, ed. *Private Science: Biotechnology and the Rise of the Molecular Sciences.* Philadelphia: University of Pennsylvania Press, 1998.

Yoxen, E. *The Gene Business: Who Should Control Biotechnology.* New York: Harper & Row, 1983.

www.thecampaign.org/

www.holisticmed.com/ge/
www.inmotionmagazine.com/geff4.html
www.nap.edu/books/0309092094/html
www.netlink.de/gen/home.html
www.nlm.nih.gov/medlineplus/ency/article/002432.htm
www.psrast.org/indexeng.htm
www.quackwatch.org/index.html

CHAPTER 5

"Altered Fish's Weak Offspring." *USA Today Magazine* 132, no. 2709 (June 2004): 15.

Brainard, Jeffrey. "FDA Warns Universities over Studies Involving Transgenic Animals." *Chronicle of Higher Education* 49, no. 39 (6 June 2003): A24.

"Cloned Horse." *Beijing Review* 48, no. 19 (12 May 2005): 8.

Dewar, Elaine. "The Second Tree: Stem Cells, Clones. Chimeras, and Quests for Immortality Down on the Pharm." *Economist* 372, no. 8393 (18 September 2004): 37–38, special section.

Everett, Jenny. "In Defense of The First Genetically Engineered Pet." *Popular Science* 264, no.2 (February 2004): 40.

Fox, Michael W. *Superpigs and Wondercorn: The Brave New World of Technology.* New York: Lyons & Burford, 1992.

Hileman, Bette. Genetic Engineering Confining Biotech Organisms." *Chemical & Engineering News* 82, no. 4 (26 January 2004): 16.

Houdebine, Marie. *Animal Transgenesis and Cloning.* Chichester, UK: John Wylie & Sons, 2003.

Kolata, Gina. *Clone: The Road to Dolly, and the Path Ahead.* New York: William Morrow and Company, 1998.

Lim, Hwa A. *Genetically Yours: Bioinforming and Biopharming, Biofarming.* Hackensack, NJ: World Scientific, 2002.

Maclean, Normal, ed. *Animals with Novel Genes.* New York: Cambridge University Press, n. d.

Reece, Richard J. *Analysis of Genes and Genomes.* West Sussex, England: John Wiley & Sons, Ltd., 2004.

Shreeve, Jamie. "The Other Stem-Cell Debate." *New York Times Magazine* 154, no. 53180, (10 April 2005): 42.

Smith, Wesley J. *Consumer's Guide to a Brave New World.* San Francisco: Encounter Books, 2004.

Westphal, Sylvia Pagán. "Growing Human Organs on The Farm." *New Scientist* 180, no. 2426–2428 (20 December 2003): 4–5.

Wilmut, Ian. "The Limits of Cloning." *NPQ: New Perspectives Quarterly* 21, no.4 (Fall 2004): 67–73.

———. *The Second Creation: Dolly and the Age of Biological Control.* New York: Farrar, Strauss and Giroux, 2000.

filebox.vt.edu/cals/cses/chagedor/quest.html
library.thinkquest.org/CR0215642/animals.html
news.nationalgeographic.com/ news/2005/01/0125_050125_chimeras.html
oslovet.veths.no/transgenics/references.html
science.gsk.com/about/animal-transgenic.
www.actionbioscience.org/biotech/margawati.html
www.acton.org/blog/?/archives/177-Celebrating-Chimeras.html
www.bio.org/animals/
www.bioethics.iastate.edu/retreat.html
www.boogieonline.com/revolution/science/pigs.html

CHAPTER 6

Begley, Sharon. "A Clue to Why Chimps, People Differ So Much Despite Similar DNA." *Wall Street Journal* 243, no. 60. Eastern edition (26 March 2004): B1.

Bethell, Tom. "The Human Genome Project: Another God That Failed." *American Spectator* 37, no. 1 (February 2004): 38–39.

Bonham, Vence L., Esther Warshauer-Baker, and Francis S. Collins. "Race and Ethnicity in the Genome Era: The Complexity of the Constructs." *American Psychologist* 60, no. 1 (January 2005): 9–15.

Carlson, Rick J. and Gary Stimeling. *The Terrible Gift: The Brave New World of Genetic Medicine.* New York: Public Affairs, 2002.

Cyranoski, David. "Japan Announces Follow-Up To Human Genome Project." *Nature* 429, no. 6990 (27 May 2004): 332.

Daiger, Stephen P. "Was the Human Genome Project Worth the Effort?" *Science* 308, no. 5720 (15 April 2005): 362–364.

"Gene-Expression Profiling: Time for Clinical Application?" *Lancet* 365, no. 9455 (15 January 2005): 199–200.

Goozner, Merrill. *The $800 Million Pill: The Truth behind the Cost of New Drugs.* Los Angeles: University of California Press, 2004.

Grossman, Dana Cook, and Heinz Valtin, eds. *Great Issues for Medicine in the Twenty-First Century.* New York: The New York Academy of Sciences, 1999.

Harris, Tina M., Roxanne Parrott, and Kelly A. Dorgan. "Talking About Human Genetics Within Religious Frameworks." *Health Communication* 16, no. 1 (2004): 105–116.

Henig, Robin Marantz. "The Genome in Black and White (and Gray)." *New York Times Magazine* 154, no. 52998 (10 October 2004): 46–51.

Holmberg, Tora. "Questioning 'The Number of The Beast': Constructions of Humanness in a Human Genome Project (HGP) Narrative." *Science as Culture* 14, no. 1 (March 2005): 23–37.

Maienschein, Jane. "Modifying Germlines—or Not?" *Nature Cell Biology* 6, no. 9 (September 2004): 797.

Marris, Emma. "Free Genome Databases Finally Defeat Celera." *Nature* 435, no. 7038 (5 May 2005): 6.

O Donnell, Christopher J. "Translating the Human Genome Project into Prevention of Myocardial Infarction and Stroke—Getting Close?" *JAMA: Journal of the American Medical Association* 293, no. 18 (11 May 2005): 2277–2279.

Pennisi, Ñelizabeth. "Can You Handle the Truth?" *Science Now* (5 May 2005): 1.

Polkinghorne, John. "The Human Genome Project." *Expository Times* 115, no. 11 (August 2004): 391–392.

Rifkin, Jeremy. *The Biotech Century: Harnessing the Gene and Remaking the World.* New York: Jeremy P. Tarcher/Putnam, 1998.

Scriver, Charles R. "The Human Genome Project Will Not Replace the Physician." *CMAJ: Canadian Medical Association Journal* 171, no. 12 (7 December 2004): 1461–1464.

Shreeve, James. *The Genome War: How Craig Venter Tried to Capture the Code of Life and Save the World.* New York: Alfred A. Knopf, 2004.

Taramelli, R., and F. Acquati. "The Human Genome Project and The Discovery of Genetic Determinants of Cancer Susceptibility." *European Journal of Cancer* 40, no. 17 (November 2004): 2537–2543.

Wickelgren, Ingrid. *The Gene Masters: How a New Breed of Scientific Entrepreneurs Raced for the Biggest Prize in Biology.* New York: Time Books, 2002.

Williams-Blangero, Sarah. "The Human Genome Project and Advances in Anthropological Genetics." *Human Biology* 76, no. 6 (December 2004): 801.

gdbwww.gdb.org/

jekyll.comm.sissa.it/articoli/art04_03_eng.htm

www.apologeticspress.org/articles/177

www.er.doe.gov/production/ober/hug_top.html

www.genome.gov/

www.kumc.edu/gec/prof/geneelsi.html

www.ornl.gov/sci/techresources/Human_Genome/home.shtml

CHAPTER 7

Avise, John C. *The Hope, Hype & Reality of Genetic Engineering: Remarkable Stories from Agriculture, Industry, Medicine, and the Environment.* New York: Oxford University Press, 2004.

Bowring, Finn. *Science, Seeds and Cyborgs: Biotechnology and the Appropriation of Life.* New York: Verso, 2003.

Boyce Nell. "Fat Chance for Meat." *U.S. News & World Report* 136, no. 6 (16 February 2004): 62–65.

Brown, Stuart F. "Soul of the New Gene Machines." *Fortune* 151, no. 9 (2 May 2005): 113–116.

Busia, Kofi. "Medical Provision in the Twenty-First Century." *Journal of Alternative & Complementary Medicine* 8, no. 2 (April 2002): 193–196.

Carey, John, Rod, and Kurtz, "Can Frayed Nerves Be Fortified?" *Business Week,* no. 3925 (21 March 2005): 83.

Carlson, Rick J. and Gary Stimeling. *The Terrible Gift: The Brave New World of Genetic Medicine*. New York: Public Affairs, 2002.

Chapman, Audrey R., and Mark S. Frankel, eds. *Designing Our Descendants: The Promises and Perils of Genetic Modifications*. Baltimore: Johns Hopkins University Press, 2003.

Coghlan, Andy. "Bad-Neighbour Genes Clash to Create Illness." *New Scientist* 178, no. 2394 (10 May 2003): 20.

Cole-Turner, Ronald. *Beyond Cloning: Religion and the Remaking of Humanity*. Harrisburg, PA: Trinity Press International, 2001.

Cook, Dana Grossman, and Heinz Valtin. *Great Issues for Medicine in the Twenty-First Century*. New York: The New York Academy of Sciences, 1999.

"The Dogs Bark, but the Caravan Still Moves On." *Economist* 375, no. 8427 (21 May 2005): 81.

Edwards, J. L., et al. "Cloning Adult Farm Animals: A Review of the Possibilities and Problems Associated with Somatic Cell Nuclear Transfer." *American Journal of Reproductive Immunology* 50 (2003).

Foster, Morris W., and Richard R. Sharp. "Will Investments in Large-Scale Prospective Cohorts and Biobanks Limit Our Ability to Discover Weaker, Less Common Genetic and Environmental Contributors to Complex Diseases?" *Environmental Health Perspectives* 113, no. 1 (January 2005): 119–122.

Fountain, Henry. "Does Science Trump All?" *New York Times* 154, no. 53229 (29 May 2005): 1, Section 4.

"Future Perfect?" *Economist* 367, no. 8330 (28 June 2003): 57.

Garwood, David. "Genetic Testing and Duty of Care." *Update* 70, no. 5 (19 May 2005): 28–31.

Gibons, Avis. "Study Results: Employer-Based Coverage of Genetic Counseling Services." *Benefits Quarterly* 20, no. 3 (2004 Third Quarter): 48–68.

Goldstein, Myrna Chandler, and Mark A. Goldstein. *Controversies in the Practice of Medicine*. Westport, CT: Greenwood Press, 2001.

Grimm, David. "Chromosome Instability Tied to Cancer." *Science Now* (12 October 2004): 1–2.

Harris, John. *Clone, Genes and Immortality: Ethics and the Genetic Revolution*. New York: Oxford University Press, 1998.

Hill, Laurie L. "The Race to Patent the Genome: Free Riders, Hold Ups, and the Future of Medical Breakthroughs." *Texas Intellectual Property Law Journal* 11, no. 2 (Winter 2003): 221–258.

Humphries, Steve. "How Genetics Will Transform CHD Risk Assessment." *Pulse* 65, no. 15 (16April 2005): 60–63.

Jones, Trevelyn E., et al. "Genetics/Medicine/Atoms and Molecules.... " *School Library Journal* 51, no. 5 (May 2005): 148.

Khoury, Muin J., et al., eds. *Genetic and Public Health in the 21st Century*. New York: Oxford University Press, 2000.

Laurie, Graeme. *Genetic Privacy: A Challenge to Medico-Legal Norms*. New York: Cambridge University Press, 2002.

Malhotra, Anil K. "Current Limitations and Future Prospects in Genetics." *Psychiatric Times* 22, no. 3 (March 2005): 18–21.

"Mapping Canine Genetics Gives Hope for Better Cancer Treatment." *DVM: The Newsmagazine of Veterinary Medicine* 35, no. 7 (July 2004): 22S.

Mayeux, Richard. "Mapping The New Frontier: Complex Genetic Disorders." *Journal of Clinical Investigation* 115, no. 6 (June 2005): 1404–1407.

McGee, Glenn. *The Perfect Baby: A Pragmatic Approach to Genetics.* Lanham, MA: Rowman & Littlefield Publishers, 1997.

Mehlman, Maxwell J. *Wondergenes: Genetic Enhancement and the Future of Society* Bloomington, IN: Indiana University Press, 2003.

Methna, John. "Revisiting the Genetics of Alcoholism." *Psychiatric Times* 21, no. 11, (November 2004): 10–12.

"Models that Take Drugs." *Economist* 375, no. 8430 (11 June 2005): 23–24.

Murphy, Timothy F. "Genetic Science and Discrimination." *Chronicle of Higher Education* 50, no. 39 (4 June 2004): B17.

Olshansky, Stuart Jay, and Bruce A. Carnes. *The Quest for Immortality: Science at the Frontiers of Aging.* New York: Norton, 2001.

Pence, Gregory E. *Re-Creating Medicine: Ethical Issues At The Frontiers Of Medicine.* Lanham, MD: Rowman & Littlefield, 2000.

Peters, Ted. *Playing God? Genetic Determinism and Human Freedom.* New York: Routledge, 2003.

Permutt, M. Alan, Jonathon Wasson, and Nancy Cox. "Genetic Epidemiology of Diabetes." *Journal of Clinical Investigation* 115, no. 6 (June 2005): 1431–1439.

Randerson, James. "A Breakthrough in Autism at Last?" *New Scientist* 184, no. 2474 (20 November 2004): 12.

Reiss, Michael, and Roger Straughan. *Improving Nature? The Science and Ethics of Genetic Engineering.* New York: Cambridge University Press, 1996.

Rowe, Steven M., Stacey Miller, and Eric J. Sorscher. "Cystic Fibrosis." *New England Journal of Medicine* 352, no. 19 (12 May 2005): 1992–2001.

Scott, Geoff. "A Century of Medical Miracles." *Current Health 2, A Weekly Reader Publication* 18, no. 5 (January 1922): 4–10.

Simons, John. "Genetic Medicine's Next Big Step." *Fortune* 151 Issue 1 (10 January 2005): 54–55.

"Spontaneous Animal Models of Human Disease." *Veterinary Record: Journal of the British Veterinary Association* 156, no. 18 (30 April 2005): 559–561.

Springen, Karen. "Using Genes as Medicine." *Newsweek* 144, no. 23 (6 December 2004): 55.

Stipp, David. "Biotech's Billion Dollar Breakthrough." *Fortune* 147, no. 10 (26 May 2003): 96–100.

Tokar, Brian, ed. *Redesigning Life?* New York: Zed Books, 2001.

Turner, Leigh. "Beware the Celebrity Bioethicist." *Chronicle of Higher Education* 50, no. 35 (7 May 2004): B18.

Turney, Jon. *Frankenstein Footsteps: Science, Genetics, and Popular Culture.* New Haven: Yale University Press, 1998.

dna.mc.vanderbilt.edu
ecoglobe.org/nz/gebiotec/health.htm
www.dnafiles.org/resources/res02.html
www.focus-on-genes.de/en/5.htm
www.geneinfo.net/
www.globalchange.com/geneticengin.htm

CHAPTER 8

Alcamo, E. I. *DNA Technology: The Awesome Skill.* New York: Harcourt Academic Press, 2001.

Aldridge, Susan. *The Thread of Life: The Story of Genes and Genetic Engineering.* New York: Cambridge University Press, 1996.

Allardice, Lisa. "Who's Your Daddy? These Days, Men Are Pulling Their Hair Out to Discover the Truth." *New Statesman* 131, no. 4589 (27 May 2002): 8.

Beckman, Mary. "Finding the DNA Needle in the Haystack." *Science Now* (30 April 2004): 2–3.

Bieber, Frederick, and David Lazer. "Guilt by Association." *New Scientist* 184, no. 2470 (23 October 2004): 20.

Brown, Jane. "The Consequences of Mistakes Are Profound." *Nursing Standard* 19, no. 36 (18 May 2005): 35.

Buss, Jessica. "DNA-Test Route to Better Marbling and Tenderness." *Farmers Weekly* 141, no. 7 (13 August 2004): 39.

Drlica, Karl A. *Double-Edged Sword.* New York: Addison-Wesley Publishing Co., 1994.

"DNA Evidence Frees 110th Prisoner." *New York Amsterdam News* 93, no. 36 (11 September 2002): 11.

"DNA Evidence Frees Falsely Accused." *USA Today Magazine* 133, no. 2718 (March 2005): 9.

"DNA's Detective Story." *Economist* 370, no. 8366 (13 March 2004): 33–35.

"From Genetic Code to Security Code." *Economist* 371, no. 8379 (12 June 2004): 14, special section.

Gigerenzer, Gerd. *Calculated Risks: How To Know When Numbers Deceive You.* New York: Simon & Schuster, 2002.

Glazer, Sarah. "Serial Killers." *CQ Researcher* 13, no. 38 (31 October 2003): 919–925.

Harrar, Sari. "DNA Fingerprinting Finds Cervical Cancer." *Prevention* 55, no. 6 (June 2003): 164.

Kafka, Tina. *DNA on Trial.* Detroit, MI: Thomson/Gale; Farmington Hills, MI: Lucent Books, 2005.

Kobilinsky, Lawrence F., Thomas F. Liotti, and Jamel Oeser-Sweat. *DNA: Forensic and Legal Applications.* Hoboken, NJ: Wiley-Interscience, 2005.

Krischner, Steve. "How 5′ to 3′ Prevents 20 to Life." *Drug Discovery & Development* 5, no.11 (December 2002): 9.

Krude, Torsten. *DNA: Changing Science and Society.* Cambridge, UK; New York: Cambridge University Press, 2004.

Lee, Henry C., and Frank Tirnady. *Blood Evidence: How DNA Is Revolutionizing The Way We Solve Crimes.* Cambridge, MA: Perseus Books Group, 2003.

Mangan, Katherine S. "Not Guilty after All." *Chronicle of Higher Education* 50, no. 40 (11 June 2004): A26–27.

Mansell, Warwick. "DNA Tests Crack the Science Problem." *Times Educational Supplement,* no. 4547 (29 August 2003): 2.

Markey, James. "New Technology and Old Police Work Solve Cold Sex Crimes." *FBI Law Enforcement Bulletin* 72, issue 9 (September 2003): 1–5.

Nossal, G. V., and Ross L. Coppel. *Reshaping Life: Key Issues in Genetic Engineering.* New York: Cambridge University Press, 2002.

"A Pandora's Box." *Economist* 365, no. 8303 (14 December 2002): 27–28.

Pattnaik, Priyabrata, and Asha Jana. "Microbial Forensics: Applications in Bioterrorism." *Environmental Forensics* 6, no. 2 (June 2005): 197–204.

Platt, Richard. *Crime Scene: The Ultimate Guide to Forensic Science.* New York: DK Publishing, 2003.

Quindlen, Anna. From "Coffee Cup To Court." *Newsweek* 139, no. 17 (29 April 2002): 80.

Randerson, James. "Forensic Clock Calls Time on Crimes." *New Scientist* 184, no. 2475 (27 November 2004): 12.

Steyn, Mark. "The Twentieth-Century Darwin." *The Atlantic Monthly* 294, no. 3 (October 2004): 206–207.

Tyre, Peg. "Reversing the Verdict." *Newsweek* 140 Issue 25 (16 December 2002): 58–59.

Urbas, Gregor. "DNA Evidence and The Forensic Process: Genetic Search And Seizure?" *Legaldate* 14, no. 3 (July 2002): 5–7.

Weinberg, Samantha. *Pointing From the Grave: A True Story of Murder and DNA.* New York: Miramax Books, Hyperion, 2003.

web.mit.edu/esgbio/www/rdna/fingerprint.html

whyfiles.org/014forensic/genetic_foren2.html

www.bergen.org/AAST/Projects/Gel/fingerprint1.htm

www.cdfd.org.in/

www.forensics.ca/links.php

www.nature.com/ng/journal/v37/n5/full/ng0505–450.html

www.ornl.gov/sci/techresources/Human_Genome/elsi/forensics.shtml

www.sciencenet.org.uk/ articlesfeatures/soc/forensics.html

CHAPTER 9

Agar, Nicholas. *Liberal Eugenics: In Defense of Human Enhancement.* Malden, MA: Blackwell Publishing, 2004.

Barbour, Ian G. "Nature, Human Nature, and God." Minneapolis, MN: Fortress Press, 2002.

Blank, Robert H., and Andrea L. Bonnicksen, eds. *Emerging Issues in Biomedical Policy,* Vol. 1. New York: Columbia University Press, 1992.

Boylan, Michael, and Kevin E. Brown. "Genetic Engineering: Science and Ethics on the New Frontier." Upper Saddle River, NJ: Prentice Hall, 2001.

Brown, Jane. "The Consequences of Mistakes Are Profound." *Nursing Standard* 19, no. 36 (18 May 2005): 35.

Buchanan, Allen E. "From Chance to Choice: Genetics and Justice." Cambridge, UK; New York: Cambridge University Press, 2000.

Burley, Justine, ed. *The Genetic Revolution and Human Rights: The Oxford Amnesty Lectures 1998.* Oxford University Press, 1999.

"Can DNA Breaks Be Repaired?" *USA Today Magazine* 133, no. 2721 (June 2005): 4–5.

Carlson, Rick J., and Gary Stimeling. *The Terrible Gift: The Brave New World of Genetic Medicine.* New York: Public Affairs, 2002.

Chapman, Audrey R., ed. *Designing Our Descendants: The Promises and Perils of Genetic Modifications.* Baltimore: Johns Hopkins University Press, 2003.Davis, Dena S. "Genetic Dilemmas: Reproductive Technology, Parental Choices, and Children's Futures." New York: Routledge, 2001.

Dolan, Kerry A. "Good Genes." *Forbes* 175, no. 12 (6 June 2005): 106–107.

Lapps, M, and R. S. Morison, eds. *Ethical and Scientific Issues Posed by Human Uses of Molecular Genetics.* New York: The New York Academy of Sciences, 2000.

Flaman, Paul. *Genetic Engineering, Christian Values, and Catholic Teaching.* New York: Paulist Press, 2002.

Gardner, Howard, Mihaly Csikszentmihalyi, and Damon William. "Good Work: When Excellence and Ethics Meet." New York: Basic Books, 2001.

Garreau, Joel. "Perfecting the Human." *Fortune* 151, no. 11 (30 May 2005): 101–105.

Garwood, David. "Genetic Testing and Duty of Care." *Update* 70, no. 5 (19 May 2005): 28–31.

Fazackerley, Anna. "Forget the Stars, the Answers Are in Your Genes." *Times Higher Education Supplement,* no. 1691 (13 May 2005): 19.

"From a Gene to a Heart Attack Drug." *Drug Discovery & Development* 8, no. 6 (June 2005): 20.

"Gender Neutral?" *Science & Spirit* 16, no. 3 (May /June 2005): 22.

"Generics Firm to Buy Rival." *Chemical & Engineering News* 83, no. 22 (30 May 2005): 18.

"Glowing Genes: A Revolution in Biotechnology." *Futurist* 39, no. 3 (May / Jun2005): 51.

Graham, Gordon. *Genes: A Philosophical Inquiry.* London; New York: Routledge, 2002.

Goozner, Merrill. *The $800 Million Pill: The Truth behind the Cost of New Drugs.* Los Angeles: University of California Press, 2004.

Gordon, Jon W. *The Science and Ethics of Engineering the Human Germ Line.* New York: Wiley-Liss, 2003.

Hanford, Jack Tyrus. "Bioethics from a Faith Perspective: Ethics in Health Care for the Twenty-First Century." New York: Haworth Pastoral Press, 2002.

Harris, John. *Clones, Genes, and Immortality: Ethics and Genetic Revolution.* New York: Oxford University Press, 1998.

Harris, Robert. "Genus Profits Soar, Despite a Slight Drop in Sales." *Farmers Weekly* Vol. 142, no. 21 (27 May 2005): 21.

Inman, Mason. "Gene Sets Humans Apart." *Science Now* (20 April 2005): 4–5.

"It's Never Too Late." *Newsweek* 145, no. 17 (25 April 2005): 76.

Justice, Scott, and Carol Singleton. "Future Perfect." *Nursing Standard* 19, no. 36 (18 May 2005). 34.

Kaplan, Jonathan Michael. "The Limits and Lies of Human Genetic Research: Dangers for Social Policy." New York: Routledge, 2000.

Kass, Leon. "Life, Liberty, and the Defense of Dignity: The Challenge for Bioethics." San Francisco: Encounter Books, 2002.

Keller, Julia C. "Straight Talk about the Gay Gene." *Science & Spirit* 16, no. 3, (May/June 2005): 21.

Kilner, John R., Rebecca D. Pentz, and Frank E. Young, eds. *Genetic Ethics: Do the Ends Justify the Genes?* Grand Rapids, MI.: William B. Eerdmans Co., 1997.

Kolata, Gina, and Sheryl Gay. "Koreans Report Ease in Cloning for Stem Cells." *New York Times* (20, May 2005).

Krimsky, Sheldon. *Science in the Private Interest: Has the Lure of Profits Corrupted Biomedical Research?* New York: Rowman & Littlefield, Inc., 2003.

Laurie, Graeme. *Genetic Privacy: A Challenge to Medico-Legal Norms.* New York: Cambridge University Press, 2002.

Lehane, Mike. "A Positive Step for Everyone's Health." *Nursing Standard* 19, no. 36 (18 May 2005): 35.

LeWine, Howard. "Tracking Family History." *Newsweek* 145, no. 17 (25 April 2005): 72.

Lewontin, Richard C. *It Ain't Necessarily So: The Dream of the Human Genome and other Illusions.* New York: New York Review Books, 2000.

Magnus, David, and Arthur L. Caplan. "Who Owns Life?" Amherst, NY: Prometheus Books, 2002.

McGee, Glenn. "Beyond Genetics: Putting the Power of DNA to Work in Your Life." New York: William Morrow and Company, 2003.

Mehlman, Maxwell J. "Wondergenes: Genetic Enhancement and the Future of Society." Bloomington, IN: Indiana University Press, 2003.

Miller, Greg. "Genetic Mixer in the Developing Brain." *Science Now* (15 June 2005): 3–4.

Minkel, J. R. "RNA to the Rescue." *Scientific American* 292, no. 6 (June 2005): 20–22.

"Models That Take Drugs." *Economist* 375, no. 8430 (11 June 2005): 23–24.

Moreira, N. "Back to Genetics." *Science News* 167, no. 24 (11 June 2005): 373.

Nossal, G.J.V., and Ross L. Coppel. *Reshaping Life: Key Issues in Genetic Engineering.* New York: Cambridge University Press, 2002.

Pence, Gregory E. "Re-Creating Medicine: Ethical Issues at the Frontiers of Medicine." Lanham, MD: Rowman & Littlefield, 2000.

Pennisi, Elizabeth. "Extra DNA Dole Makes for Faithful Vole." *Science Now* (9 June 2005): 2–3.

Rollin, Bernard E. *The Frankenstein Syndrome: Ethical and Social Issues in the Genetic Engineering of Animals.* New York: Cambridge University Press, 1995.

Rothman, Barbara Katz. *The Book of Life: A Personal and Ethical Guide to Race, Normality, and the Implications of the Human Genome Project.* Boston: Beacon Press, 2001.

Seppa, Nathan. "Mitochondria Genes May Influence Cancer Risk." *Science News* 167, no. 19 (7 May 2005): 302.

Shreeve, James. *The Genome War: How Craig Venter Tried to Capture the Code of Life and Save the World.* New York: Alfred A. Knopf, 2004.

———. "The Other Stem-Cell Debate," *New York Times* (10 April 2005).

"Slacker Genes Wait Their Chance." *New Scientist* 186, no. 2498 (7 May 2005): 18.

Smith, Wesley J. *Consumer's Guide to a Brave New World.* San Francisco: Encounter Books, 2004.

Spallone, Pat. *Generation Games: Genetic Engineering and the Future of Our Lives.* Philadelphia: Temple University Press, 1992.

Stipp, David. "How Disease Evolves." *Fortune* 151, no. 10, (16 May 2005): 53–54.

Stock, Gregory. *Redesigning Humans: Our Inevitable Genetic Future.* New York: Houghton Mifflin Co., 2002, 178.

Sulston, John, and Georgina Ferry. "The Common Thread: A Story of Science, Politics, Ethics, and the Human Genome." Washington, DC: Joseph Henry Press, 2002.

Thistlethwaite, Susan Brooks. "Adam, Eve, and the Genome: The Human Genome Project and Theology." Minneapolis, MN: Fortress Press, 2003.

Tokar, Brian. "Redesigning Life? The Worldwide Challenge to Genetic Engineering." New York: Zed Books, 2001.

Williams, Patricia J. "Genetically Speaking." *Nation* 280, no. 24 (20 June 2005): 10.

Wright, Stephen. "It's All in the Genes." *Nursing Standard* 19, no. 37 (25 May 2005): 32–33.

Yount, Lisa. "The Ethics of Genetic Engineering." San Diego, CA: Greenhaven Press, 2002.

Index

About the Author

MARK Y. HERRING is Dean of Library Services at Winthrop University. He has worked in librarianship for more than two decades. He has written numerous books, articles, and scores of reviews for magazines and journals both in and out of the library profession, including *The Pro-Life/Choice Debate* (Greenwood, 2003).